The Political Economy of Reform in Post-Mao China

© Copyright 1985 by the President and Fellows of Harvard College

The Council on East Asian Studies at Harvard University publishes a monograph series and, through the Fairbank Center for East Asian Research and the Japan Institute, administers research projects designed to further scholarly understanding of China, Japan, Korea, Vietnam, Inner Asia, and adjacent areas. Publication of this volume has been assisted by a grant from the Shell Companies Foundation.

Library of Congress Cataloging in Publication Data

Main entry under title:
The Political economy of reform in post-Mao China.

   (Harvard Contemporary China Series ; 2)
   Includes index.
   1. Agriculture and state—China. 2. Industry and state—China. I. Perry, Elizabeth J.
II. Wong, Christine, 1950-   . III. Series.
HD2098.P64   1985   338.951   85-5529
ISBN 0-674-68590-3

The Political Economy of Reform in Post-Mao China

Harvard Contemporary China Series: 2

edited by
ELIZABETH J. PERRY
and CHRISTINE WONG

Published by
THE COUNCIL ON EAST ASIAN STUDIES / HARVARD UNIVERSITY
Distributed by the Harvard University Press
Cambridge (Massachusetts) and London 1985

CONTRIBUTORS

KATHLEEN HARTFORD is Associate Professor of Political Science at the University of Massachusetts in Boston. Her monograph on the recent changes in Chinese agricultural policy will be published by the Columbia East Asian Institute's Occasional Papers series. She is currently completing a book on the Chinese Revolution and working on a comparative study of socialist agriculture.

JOYCE K. KALLGREN is Professor of Political Science (Davis campus) and Chair of the Center for Chinese Studies (Berkeley campus) at the University of California. She is now engaged in a study of Chinese public policy with respect to the disabled and disadvantaged.

RICHARD J. LATHAM is Associate Professor of Political Science and Director of Comparative and Area Studies at the United States Air Force Academy. He is the author of several articles on Chinese science and technology and recently completed a doctoral dissertation at the University of Washington on the impact of recent reforms on the Chinese Communist Party.

BARRY NAUGHTON is Assistant Professor of Economics at the University of Oregon. His current research interests are the industrial planning and management system in China and urban incomes and living standards. Professor Naughton spent a year in China hosted by Wuhan University.

ELIZABETH J. PERRY teaches Chinese politics in the Jackson School of International Studies at the University of Washington. Author of *Rebels and Revolutionaries in North China* and editor of *Chinese Perspectives on the Nien Rebellion*, she is currently working on a study of rural collective action in socialist China.

LOUIS PUTTERMAN is Associate Professor of Economics at Brown University. His publications include articles on the economic theory of work organization and incentives and a forthcoming book on agricultural cooperation entitled *Peasants, Collectives, and Choice*.

SUSAN L. SHIRK is Associate Professor of Political Science at the University of California, San Diego. She is the author of *Competitive Comrades* and is now at work on a study of industrial reform in contemporary China.

TERRY SICULAR received the PhD from Yale in 1983 and is now a member of the Food Research Institute at Stanford University. She has written several papers on Chinese agricultural planning and price policy.

S. LEE TRAVERS works at a commodities trading firm in New York and is the author of articles on rural income in China.

CHRISTINE WONG teaches economics at Mount Holyoke College. During the 1984–1985 academic year, she is Visiting Assistant Professor of Economics and a Postdoctoral Fellow at the Center for Chinese Studies, University of California, Berkeley, where she is completing a study of rural industrialization in China.

CONTENTS

Introduction: The Political Economy of Reform in Post-Mao China: Causes, Content, and Consequences 1
ELIZABETH J. PERRY and CHRISTINE WONG

PART ONE AGRICULTURE

1 *Socialist Agriculture is Dead; Long Live Socialist Agriculture! Organizational Transformations in Rural China* 31
KATHLEEN HARTFORD

2 *The Restoration of the Peasant Household as Farm Production Unit in China: Some Incentive Theoretic Analysis* 63
LOUIS PUTTERMAN

3 *Rural Marketing and Exchange in the Wake of Recent Reforms* 83
TERRY SICULAR

4 *Getting Rich through Diligence: Peasant Income after the Reforms* 111
S. LEE TRAVERS

5 Politics, Welfare, and Change: The Single-Child Family in China 131
JOYCE K. KALLGREN

6 The Implications of Rural Reforms for Grass-Roots Cadres 157
RICHARD J. LATHAM

7 Rural Collective Violence: The Fruits of Recent Reforms 175
ELIZABETH J. PERRY

PART TWO INDUSTRY

8 The Politics of Industrial Reform 195
SUSAN L. SHIRK

9 False Starts and Second Wind: Financial Reforms in China's Industrial System 223
BARRY NAUGHTON

10 Material Allocation and Decentralization: Impact of the Local Sector on Industrial Reform 253
CHRISTINE WONG

Notes 281

Index 320

TABLES

1. Incidence of Production Responsibility Systems — 40
2. Grain Marketing and Resales in Rural Areas, 1952–1982 — 90
3. Edible Vegetable Oil Production and Marketing, 1953–1982 — 93
4. Cotton Production, Marketing, and Retention, 1952–1982 — 95
5. Price Indexes for Agricultural Procurement and for Industrial Retail Sales in Rural Areas, 1952–1982 — 98
6. Quota Procurement Prices for Agricultural Products, 1952–1979 — 100
7. Quota and Above-Quota Procurement Price Indexes for First-Category Agricultural Products, 1978 and 1982 — 101
8. Production and Procurement of First-Category Products — 102
9. Rural Markets: Number and Total Value of Transactions — 105
10. National Average Rural Market Prices and State Above-Quota Prices for Rice, Wheat, and Vegetable Oils, 1980 and 1981 — 106

11. Composition of Peasant Income and Year-to-Year Increases, 1978–1981 — 113
12. Estimated Source of Growth in Collective Income, 1979–1981 — 115
13. Year-to-Year Changes in Collective Income and Costs, 1977–1981 — 122
14. Per Capita Retail Sales of Consumer Goods, 1978–1981 — 126
15. Per Capita Personal Savings, 1978–1981 — 127
16. Chinese Population Projections for 2000 and 2080 — 140
17. Geographic and Demographic Data for Production Teams of Respondents, Guangdong, 1981 — 162
18. Geographic and Demographic Data for Production Teams of Respondents, 1981 — 163
19. The Impact of Agricultural Reforms on Rural Leadership Elites, Guangdong, 1980–1981 — 164
20. The Impact of Agricultural Reforms on Rural Leadership Elites, 1980–1981 — 165
21. Central and Local Fixed Investment, 1977–1982 — 229
22. Total Local Funding Sources, 1980, 1981, 1982 — 233
23. Profit Retained by State-Run Enterprises, 1978–1982 — 237
24. Number of Producer Goods under State Allocation, 1953–1981 — 259
25. Local Control of Category I and II Materials, 1978 — 262
26. Local Control of Category I and II Materials (as Percent of Total Output) — 267
27. Profit and Industrial-Commercial Tax Rates for Selected Industries — 270
28. Selected Industries with High Industrial-Commerical Tax Rates — 273

ACKNOWLEDGMENTS

Since the death of Chairman Mao, China has witnessed head-spinning change as new economic and political policies have ushered in a series of sweeping reforms which seemed unimaginable only a few years ago. The papers that comprise this volume represent an effort to explore the sources and significance of the reforms. The rapidity with which new developments are taking place threatens any study of contemporary China with early obsolescence. We hope, however, that a respectable life span for this volume will be ensured by the scope of the papers—all of which address central issues of Chinese political economy that are likely to outlast short-term policy shifts. We further hope that the volume's initial publication in soft-bound format (for which we thank the Harvard Council on East Asian Studies Publications Committee) will encourage its adoption for room use.

This volume grew out of the workshop on "Recent Reforms in China: Economic, Political, and Social Implications" held at Harvard University in April 1983. As co-organizers of the workshop, we should like to thank the individuals and institutions that facilitated its success—especially Merle Goldman, Sylvia Appleton, Deborah Knosp, Philip Kuhn, Patrick Maddox, and Dwight Perkins; the Fairbank Center for East Asian Research, the New England China Seminar, and the Harvard Institute for International Development. A debt of gratitude is owed to the workshop participants and other contributors who helped to turn the conference papers into this volume. Finally, support for completion of the volume was also provided by Mount Holyoke College and the University of Washington's Jackson School of International Studies.

Elizabeth J. Perry
and
Christine Wong

ABBREVIATIONS USED IN TEXT AND NOTES

AFP	Agence France Press
AIC	agriculture-industry-commerce integrated company
BBC	British Broadcasting Corporation
CCP	Chinese Communist Party
CR	Cultural Revolution
CRPSMA	Joint Publications Research Service, China Report: Political, Sociological and Military Affairs
CZ	*Caizheng* (Finance)
FBIS	Foreign Broadcasts Information Service
FYP	Five-year plan
ICM	*Inside China Mainland*
JETRO	Japan External Trade Organization
JJGL	*Jingji guanli* (Economic management)
JJNJ	*Zhongguo jingji nianjian* (Almanac of China's economy)
JJYJ	*Jingji yanjiu* (Economic research)
JPRS	Joint Publication Research Service
MMB	Ministry of Machine Building

MOFERT	Ministry of Foreign Economic Relations and Trade
NCNA	New China News Agency
PLA	People's Liberation Army
PRC	People's Republic of China
PRS	Production responsibility system
RR	Replacement and reconstruction
RRS	rural responsibility system
SCF	Single-Child Family
SEC	State Economic Commission
SEZ	Special Economic Zone(s)
SPC	State Planning Commission
SSB	State Statistical Bureau

The Political Economy of Reform in Post-Mao China

INTRODUCTION

The Political Economy of Reform in Post-Mao China: Causes, Content, and Consequences

ELIZABETH J. PERRY *and* CHRISTINE WONG

The major reforms announced in December 1978 at the Third Plenum of the Eleventh Congress of the Chinese Communist Party mark a watershed in the political economy of contemporary China. Dramatic changes in both agriculture and industry seemed to spell a near total repudiation of the Maoist "road to socialism" which had been constructed over the preceding three decades. Once regarded as the epitome of radical communitarianism, China was suddenly pictured by many outside observers as a budding capitalist society alive with the spirit of entrepreneurial individualism. What precipitated this abrupt change in development strategy? What are the objectives and implications of the new reform measures?

Although it is too early to offer firm conclusions on the

extent to which the reforms can or will attain their goals, a preliminary assessment is now possible. The ten chapters this volume comprises—written by five economists and five political scientists, all specialists on China—provide analyses of the results to date, some six years after the Third Plenum. Because of the quite different nature of the reforms in agriculture and industry, the volume is divided into two sections. Part One delineates the content of the agricultural reforms and considers their impact on production, marketing, peasant income, family planning, local leadership, and rural violence. Part Two analyzes the evolution of industrial reforms and their impact on political conflict, resource allocation, investment, material and financial flows, industrial structure, and composition of output.

One theme that emerges clearly from these studies—despite their differences in emphasis and interpretation—is the interconnection of economics and politics. The origins of the reforms are found in the combination of poor economic performance and disruptive political upheavals resulting from the Cultural Revolution and its aftermath. Similarly, the future of the reforms will depend upon developments in both the economy and the political arena—developments that are, in turn, closely related. This introductory chapter will attempt to sketch some of the ways in which politics and economics interact in the causes, content, and consequences of China's post-Mao reforms.

BACKGROUND TO REFORMS

Widespread popular discontent with economic performance during the "Cultural Revolution decade" (1966–1976) played a major role in precipitating the recent reforms. In the rural sector, slow growth in agricultural output, combined with rapid population growth and lack of opportunities for off-farm employment, resulted in a low per capita income growth of only 0.5 percent per annum; from 103 yuan in 1957 to only 113 yuan in 1977 (see Lee Travers's chapter below). Aside from the adverse effect of this slow income growth on peasant incentives and morale, the stagnation in agricultural production led to a decline in marketing rate, which, in turn, posed growing problems for urban food supplies. Shortages of non-grain food products brought a proliferation of local rationing systems through the mid-1970s; pork rationing was introduced in Shanghai in July 1976 for the first time since the post-Great Leap Forward crisis of the early 1960s. Along with edible oils, sugar, and cotton, grain imports rose throughout the 1970s, and it is estimated that, by the mid-1970s, over one-third of urban grain consumption came from imports. In the face of the government's stated objective of achieving self-sufficiency in food production, the Cultural Revolution agricultural policy must be judged a failure.

This failure was all the more vexing, given China's impressive record in popularizing new seed technology during the Cultural Revolution (CR) period. In 1964, China began to introduce high-yielding, dwarf varieties of rice seeds developed independently of the concurrent efforts at the International Rice Research Institute. This breakthrough was accompanied by massive investments in the chemical-fertilizer and farm-machinery industries. With the development of domestic production capability (notably in small-scale plants), chemical fertilizer output rose rapidly from 0.65 million tons (100 percent nutrient content) in 1963 to 7.23 million tons in 1977.[1] Along with water conservancy projects undertaken in the 1950s, these increased supplies of modern inputs allowed rapid dissemination of high-yielding rice strains; by 1977, over 80 percent of total sown acreage in rice was planted with such varieties.[2] Given this technological progress, the stagnation in agricultural output points to serious organizational and incentive problems.

In industry, although gross output grew at an impressive rate of nearly 10 percent per annum during 1957–1979, this performance was undermined by problems of inefficiency and poor coordination. Excessive investment in heavy industry and insufficient attention to the development of supporting industries and infrastructure led to mismatches between supply and demand. According to one informed report, shortages of fuel, electric power, and transport facilities caused 20–30 percent of industrial capacity to go unutilized during 1975–1977, costing an estimated 75 billion yuan in output foregone.[3] Yet, during the same period, investment continued apace, mostly in creating duplicative production capacity rather than in the bottleneck energy and transport sectors.

The intense competition for inputs, chronic shortages, and unutilized capacity caused dramatic deterioration in a number of performance indicators. According to a report published in the *People's Daily* by two senior economists at the Chinese Academy of Social Sciences, the national income produced per 100 yuan of fixed assets averaged 34 yuan during the 1976-1979 period, compared with 52 yuan during the First Five Year Plan.[4] Over one-third of all state-owned enterprises were running at a loss in 1976.[5] In 1978, 43 percent of quality and 55 percent of consumption norms in industry could not meet the best levels set in the 1960s.[6] Along with the inefficient use of resources, the high rate of accumulation maintained during the 1970-1976 period—31-34 percent of national income, compared with 20-25 percent during the First Five Year Plan—also served to depress current consumption.

In the urban sector, wages for state workers were virtually frozen from 1963 to 1977, with only a partial adjustment in 1971-1972 for those in the bottom two grades of the pay scale. With bonus pay eliminated during the CR, and without periodic promotions for seniority and skill acquisition, many workers suffered a decline in real income. As young workers were added at the bottom of the wage ladder, the average wage in the state sector declined from a peak of 741 yuan in 1964 to 632 yuan in 1977. Although urban living standards actually rose during the period—with rising *family* incomes due to increased labor participation—stagnant wage rates and a breakdown in work-place discipline produced extremely low worker morale and high absenteeism.[7]

The deterioration in economic performance clearly pointed to a need for some type of reform. But the history of reform movements in other countries shows that implementation of reform does not follow automatically upon the heels of economic difficulties. In terms of politics, reforms demand an extraordinary degree of acumen and finesse. As Samuel Huntington has noted, "Reform is rare if only because the political talents necessary to make it a reality are rare."[8] It is thus of some interest to explore the political conditions that have made possible the dramatic reform effort currently underway in China.

While it is impossible to know with certainty just why the Chinese leadership felt compelled to effect such a radical break with Maoist precedents, several possible explanations come to mind. The most obvious involves the bitter experience of the Cultural Revolution, when most of the current leaders suffered years of public humiliation, if not physical abuse, at the hands of Maoist accusers. One result of this experience, probably accentuated by Deng's competition with Hua Guofeng in the mid-1970s, was the survivors' demand for a dramatic departure from the legacy of Maoism. The People's Communes—instituted during the intitial euphoria of the Great Leap Forward—were seen as a central part of that legacy, as was even the practice of collectivized agriculture, whose "high tide" had convinced Mao of the feasibility of moving full steam ahead to communes.

It is commonplace in the study of contemporary China to remark upon the abrupt swings in policy which occur when the locus of decision-making authority shifts from one set of leaders to another.[9] Moreover, these swings of the political pendulum have, over the years, become increasingly radical—as each group attempts to implement programs that will set the country on an irreversible course. The frantic pace of Mao's Cultural Revolution was in part a reflection of the aging Chairman's impatience and desire to ensure successors loyal to his ideals. Similarly, 80-year-old Deng Xiaoping is possessed by an urgency to propel China well along the road toward the Four Modernizations before he passes from the scene. Just as the Cultural Revolution entailed a "leftist" swing of dimensions which far exceeded its precursors (for example, the Hundred Flowers or the Great Leap), so the current "rightist" reversal goes much farther than earlier precedents.

But the recent reforms are not simply a response to domestic political imperatives. They also reflect China's new perception of the international climate and her role in the world economy. The many examples of developing countries (including the Chinese-dominated economies of Hong Kong, Taiwan, and Singapore) that successfully tapped the world market for growth during the 1960s and early 1970s undoubtedly helped motivate

China to abandon its autarkic stance. Although the "great leap outward" into the international system was initiated during the "Gang of Four" era of the early 1970s, it was obviously promoted by the tireless efforts of then Premier Zhou Enlai and (from 1973 to 1976) his close associate Deng Xiaoping. Symbolized by subsequent normalization of relations with both Japan and the United States, this new international role facilitated a dramatic change in economic strategy. No longer would local areas be required to strive for industrial and agricultural self-sufficiency in defense against a hostile world. Instead, industry could be revamped and modernized along lines of regional specialization and division of labor, realizing economies of scale and adopting foreign technology as appropriate. In agriculture, peasants were encouraged to grow the crops best suited to local conditions—including commercial crops (like cotton) to supply the development of light industry for an expanding export market.

The package of reforms enunciated in December 1978 thus had both economic and political precipitants. And the goals of the reforms include both economic and political objectives. In addition to raising productivity and improving the standard of living, the new measures are also designed to prevent the possibility of another Cultural Revolution by rectifying Party ranks, strengthening the legal system, and introducing more democratic procedures.

The seeds of the current drive for reform were planted during the tumultuous Cultural Revolution decade, but did not germinate until the death of Mao Zedong and arrest of the Gang of Four in the autumn of 1976. Since Chairman Mao and his radical sidekicks had been responsible for many of the excesses of the Cultural Revolution, their demise opened the door to a change of course. As the recent histories of other socialist states make clear, leadership succession is often accompanied by a reformist thrust on the part of the new regime,[10] especially when economic problems can be traced to the visible hand of the old regime.

The chief engineer and initial beneficiary of the Gang of Four's arrest was Mao's alleged anointed successor, Hua Guofeng. Hailing from Mao's native province and bearing a marked physical resemblance to his predecessor, Chairman Hua would seem to have had good reason for maintaining some continuity with past policies. Inasmuch as Mao and the Cultural Revolution had made possible Hua's sudden rise to prominence, it appeared doubtful that the new Chairman could break with the basic tenets of Maoism. As things turned out, however, Hua Guofeng began to advocate significant changes after only a few months in office. At a series of conferences in early 1977, Hua stressed the need for rapid economic modernization, with improvement in the standard of living emphasized as a key objective.[11]

In retrospect, it becomes clear that Hua Guofeng was Mao's choice for successor because he represented a compromise between the "ultra-leftism" of the Gang of Four and the "capitalist-road" tendencies of Deng Xiaoping. In the final days of Mao's life, as competing factions jockeyed for position, the aging dialectician apparently decided to settle on a successor who promised to combine attributes of both camps. While Hua's past would seem to have put him closer to the radicals, the political context of the day in fact argued for a different path. With Deng Xiaoping temporarily eliminated from the rivalry by his dismissal in the spring of 1976, Hua soon found himself struggling against the Gang of Four to maintain his grip on power. It may have been this competition that prompted Hua's daring arrest of the Gang in October 1976. But, once the radicals were behind bars, Hua faced the tricky problem of distinguishing himself from his now discredited rivals.

As Valerie Bunce has shown with respect to both the Soviet Union and the United States, new leaders commonly launch reform initiatives to shed the stigma of past associations.[12] Leadership succession in socialist systems, because of its infrequency, is particularly likely to generate policy innovation. New leaders are commonly confronted by a backlog of problems accumulated over the years by their predecessors. (This was certainly true in Hua's case, for Mao had served as Party Chairman for some four decades before his death.) To prove themselves as bold and legitimate rulers, successors attempt to chart a new course more responsive to popular needs. Their reforms usually evidence a strong concern for welfare provisions; proposals for a higher standard of living and for more consumer goods are common features.

While announcement of a new, reformist program is not difficult, carrying out such a project requires considerable skill. And, as history would quickly show, Hua Guofeng was not up to the task. Instead, the mission would fall to his erstwhile rival, Deng Xiaoping. Almost immediately after his victory over the Gang of Four, Hua faced strong pressure from senior Party statesmen to rehabilitate Deng. Since Hua, as former Minister of Public Security, had been responsible for Deng's dismissal six months earlier, the issue of rehabilitation was somewhat embarrassing. An even greater concern was Deng's extraordinary political finesse. A veteran of the Long March and long-time aide to Premier Zhou Enlai, Deng was known as a tough, determined, decisive, and persuasive politician. These qualities endeared Deng to the elder Party statesmen (men such as Ye Jianying and Li Xiannian), whose advice Hua had apparently followed in getting rid of the Gang of Four and whose support could not be forfeited if he was to survive. Reluctantly, Hua extended an olive branch to his formidable rival.

After a series of delicate maneuvers, Deng Xiaoping was officially restored to his previous Party and government positions at a Central Committee meeting in July 1977. As Vice-Premier, Deng was well placed to vie with Hua Guofeng for the title of reformer *par excellence.* Although it is unclear to just what extent the relationship between the two men was marked by conflict or cooperation during this period, the outcome was certainly a boost for the spirit of reform. Hua and Deng both championed the project of modernizing China by the year 2000, a project that had been enunciated by the late Premier Zhou in a January 1975 speech to the Fourth National People's Congress. On that occasion, Zhou elaborated the theme of Four Modernizations (agriculture, industry, science-technology, and national defense) which he had first introduced more than a decade before at the Third National People's Congress.

At the Fifth National People's Congress in February 1978, Hua Guofeng announced an ambitious ten-year modernization program intended to accelerate China's progress toward achieving the Four Modernizations by 1985. In the Maoist tradition of stressing revolutionary fervor and state initiative, the plan called for substantial investment in new construction projects and very high growth rates in both industry and agriculture. Flushed with enthusiasm, government officials hastily approved some 100,000 construction projects and negotiated scores of foreign contracts (such as the ill-fated Baoshan steel complex) in an effort to mobilize the capital and technology necessary for rapid economic development.

Although announced in early 1978, many components of the plan had actually already been implemented in 1977. Coming after a decade of accumulated problems and working with an unreformed economic and political structure, the Hua program quickly unleashed a "mini Great Leap" in investment that exacerbated sectoral imbalances and revealed glaring weaknesses in the central planning apparatus. In June 1979, Hua was forced to announce a three-year period of "readjustment." The grandiose ten-year plan unveiled the previous year was now quietly scrapped under the sober realization that problems with energy, transportation, foreign credit, and the like made impossible the forecasted leap in economic growth rates. The retrenchment signaled an important change in direction and an admission that the Maoist development strategy was in need of substantial alteration.

It can be said that genuine reform began with the demise of Hua Guofeng's ten-year plan. Hua's program, rather than representing a true break from CR policies, was an attempt to correct certain excesses of ultra-leftism: the regional self-sufficiency drives in agriculture and industry, the suppression of material incentives (workers' bonuses were reintroduced on an experimental basis in mid-1978), the autarkic stance vis-à-vis foreign trade, and so on. The 1977–1978 period was marked by attempts to strengthen centralized control over financial and material resources. Industrial policies called for renewed efforts to improve standardization of products and production techniques, reasserting ministerial control over large-scale enterprises, and extending supervision over small-scale enterprises. Economic and administrative bureaucracies were to be beefed up to strengthen their capacities for managing the economy and fine-tuning policies. The failure of the ambitious growth policies of the period hastened Hua's demise and helped to further discredit centralized planning and to legitimize market reforms.

CONTENT OF THE REFORMS

The most daring reform measures implemented since the Third Plenum have occurred in the countryside, home to some 80 percent of China's one billion inhabitants. The single most important agricultural reform is the introduction of production responsibility systems, a program which has drastically altered the organization of production and the distribution of output in the countryside. Alongside this radical transformation in rural organization have come changes in overall planning which alter the relationship between agriculture and the rest of the economy. Reforms in planning aim to stimulate production and procurement by reducing the degree of central control and providing more incentives to producers. Specifically, the changes in planning have included adjustments in procurement quotas, price reform, and a relaxation of marketing restrictions.

If these constitute the main contours of the agricultural reforms (described in detail in the following chapters), what have been their initial results? In terms of productivity, procurement, and peasant income, some significant progress has been achieved. Reduction of state procurement quotas and liberalization of trade ended enforced local self-sufficiency in the post-Mao period, and many localities have reverted to traditional cropping patterns. Terry Sicular notes that increased regional specialization and exchange account for a substantial part of the recent growth in productivity: Output of key commercial crops such as cotton and edible oil grew by an impressive 13.5 percent and 22.7 percent per annum, respectively, from 1978 to 1982, while grain output grew at a respectable 2.4 percent.

From 1978 to 1982, quota procurement prices for the major crops of grain, cotton, and edible oil were raised by 26, 20, and 24 percent, respectively, while above-quota prices (inclusive of bonuses) were raised by averages of 45, 56, and 43 percent. Reduced procurement quotas provided an added incentive for peasants to increase production and sales. At present, four types of prices are applied to agricultural products: quota, above-quota, negotiated, and free-market prices. With the above-quota price for grain set at 50 percent higher than the quota price, an increase in the proportion of grain procured at above-quota prices raises the average price of procurement. Indeed, the regime's success in raising grain procurement from 46.5 million tons in 1978 to 54.55 million tons in 1981 is partly due to its willingness to change the procurement mix. While state procurement remained at a constant 16 percent of total grain output from 1978 to 1981, the portion of procurement from base quota and taxes fell from over 60 to 40 percent (Sicular). In short, the regime could claim substantial progress in the areas of both production and procurement.

Along with these achievements went an improvement in the peasants' standard of living. Lee Travers estimates that, from 1978 to 1981, real per capita rural income grew at an annual rate of 11.4 percent. Much of the improvement can be attributed to the liberalization of sideline production and rural marketing. Following experiments in Sichuan that showed private-plot productivity to be much higher than that of collective land, the limit of 5-7 percent of cultivated land in private plots was raised to 15 percent. Nationally, the amount of cultivated land devoted to private plots rose from 5.7 percent in 1978 to 7.5 percent in 1981. With increased opportunities for marketing and higher free market prices, income from household sidelines has been the fastest growing component of peasant income, rising from 37.7 billion yuan in 1978 to 67.2 billion yuan in 1981. If this trend continues, it will soon overtake income from collective agriculture as the largest source of peasant income; by 1981 it accounted for 40.5 percent of total rural income, compared with 48.7 percent from collective agriculture.

Reforms in the industrial sector have been less radical than the new agricultural policies, but they have nevertheless brought significant changes to the planning and organization of production and investment, as well as substantial improvement in worker remuneration and welfare. The centerpiece of industrial reforms is enterprise profit retention, a program that has gone through a number of metamorphoses, from the Sichuan experiment in 1978-1979 to profit contract systems in 1981-1982 to tax-for-profit schemes in 1983. In order to make enterprises truly autonomous units bearing significant responsibility for their operations, a number of related reforms were introduced, including the use of repayable bank loans to replace state grants for investment financing, collection of interest charges on enterprise fixed assets and working capital, retention of depreciation funds by enterprises, permission for enterprises to sell extra-plan output through market channels, limited use of flexible prices, and decentralization of material allocations. Wages in the state sector were raised in three rounds during 1978–1980, and bonuses were reinstated in 1978–1979, tied directly to enterprise profit and plan fulfillment. Together, the reforms have led to an unprecedented decentralization of control over financial and material resources, substantially altering the relationship between the state and the enterprise, and between market and plan in resource allocation.

Experiments in enterprise profit retention began with six pilot firms in Sichuan Province in October 1978. This was followed in July 1979 by nationwide implementation under the National Pilot Program. Concern with stabilizing state revenues led the Ministry of Finance to push for limiting participation to no more than 1,600 enterprises. Once introduced, however, the reform began to gather momentum. By June 1980, 6,600 of the largest and most profitable enterprises were enrolled. Although comprising only 11 percent of the total of state-owned enterprises, they accounted for 60 percent of the output and 70 percent of the profits of the in-budget industrial sector. In April 1981, the pace accelerated even farther with the introduction of "economic responsibility systems," which turned profit-sharing into a profit contract system under which enterprises negotiated annual profit-remittance quotas, retaining most or all above-quota profits. By year-end, 80 percent of all state-owned enterprises were under profit contracting. The continuing decline of enterprise profit remittances under this scheme and the difficulties of setting realistic profit contracts resulted in yet another metamorphosis—the tax-for-profit scheme introduced in 1983. Eventually, this system is to replace all profit-retention schemes and reduce state-enterprise financial interaction to that of taxation. In addition to stabilizing state revenues, it is hoped that this system will restore greater objectivity in determining enterprise incentives and fairness to the financial system (see the chapter by Barry Naughton).

Unlike rural policies, industrial reforms have produced little clear-cut evidence of success in increasing efficiency or reorienting production the better to match demand. Given the complex interaction among the web of reform measures affecting industrial performance, there is some disagreement among observers over efficacy of the reforms to date and their future prospects. An argument for some improved production efficiency at the enterprise level is offered by William Byrd, who estimates that the ratio of current inputs to gross value of output in industry (adjusted for price changes) fell by some 4-7 percent during 1978-1981, indicating that enterprises became more frugal in their use of materials in response to incentives provided by profit-retention schemes. However, this assessment is weakened by his data showing stagnant factor productivities.[13]

There is, in addition, general agreement that macroeconomic efficiency remains low and has even declined in the reform period, thereby negating any marginal gains in microeconomic efficiency. The chief culprit in poor macroeconomic performance is excessive and inappropriate investment. Despite the government's repeated calls to curtail investment so as to concentrate resources on improved utilization of existing plant and equipment, investment in fixed assets increased from 66.9 billion yuan in 1978 to 84.6 billion yuan in 1982. All the increase has come from local investment, which more than doubled (from 22.5 billion yuan to 53.6 billion yuan) during this period. This investment is financed out of the vastly increased resources that profit-retention and depreciation allowances have bestowed on enterprises, along with new funds transferred to local government revenues by fiscal decentralization (Naughton).

The chapters by Barry Naughton and Christine Wong challenge the assumption that the economic maladies of the Cultural Revolution decade stemmed from over-centralization. Instead, they show that the Chinese economy had in fact become highly decentralized during the Cultural Revolution period. Naughton traces the decentralization of financial control to 1967, with the transfer of depreciation allowances to the enterprises. Looking at the material allocation system, Wong argues that the rapid growth of local industry during the Cultural Revolution had put substantial resources in local hands, thereby strengthening local governments vis-à-vis Beijing. By the end of the Cultural Revolution, major portions of key materials were controlled by local governments, and material allocation was increasingly conducted on a regional basis.

Fiscal decentralization and market reforms since 1978 have further enhanced the power of local governments. Changes in the material-allocation system have greatly increased the quantities of materials traded outside state plan channels. Revenue-sharing arrangements between the central and provincial governments and between provincial and county governments have transformed each level into a relatively independent economic agent seeking to maximize revenues, often by investing in new projects to expand its revenue base. Similarly, progressive simplification of enterprise incentive schemes has increased local government intervention in enterprise operations. (Mis)guided by a price and tax structure that yields across industries unequal rates of return that are uncorrelated with scarcity values or demand, local investment has been concentrated in a few processing and assembly industries. This leads to "blind and duplicative" projects which contribute to low capital efficiency and exacerbate problems of regionalism. To be sure, *some* restructuring of production to meet consumer demands has taken place over the past few years, although demand still far outstrips available supplies for many products. And, predictably, some of the "blind and duplicative" problem in heavy industry has simply been shifted over to the light industrial sector.

In neither agriculture nor industry have reforms been implemented as a consistent package in line with some clear blueprint for decentralized, market socialism. Rather, reform measures were initiated in a piecemeal, incremental fashion—often in reaction to specific problems. Typically, reforms began cautiously with small-scale experiments in carefully selected areas: Agricultural production responsibility systems were first tested in Anhui province and enterprise-retention schemes began with the Sichuan pilot firms. The design of these programs often was altered to accommodate or ameliorate problems that surfaced in the course of wider implementation. Naughton's discussion of financial reforms in the industrial system provides an instructive example of this process.

Despite the careful, tentative way in which many of the reforms were initially tested, their subsequent approval by the central leadership often resulted in a frantic rush for rapid implementation with little regard for local conditions. Kathleen Hartford points to considerable resistance to the agricultural production responsibility systems in some areas. Although the regime claims that new measures have been introduced democratically, Hartford notes that the initial stress on voluntarism was quickly superseded by a strong push for compliance. Once again the tactic of "cutting with a single stroke of the knife" (the universal application of central policy without room for local variation) was being employed by a bureaucracy whose operating procedure is notoriously insensitive to regional differences.

At first glance, the various components of the recent economic reforms would seem internally consistent and mutually reinforcing. For example, reforms granting enterprises greater autonomy were complemented by measures that decentralized control over financial and material resources, allowing enterprises to obtain resources needed to substantiate their production and investment decisions. Similarly, agricultural reforms that gave production units greater flexibility over cropping patterns were accompanied by the lifting of many restrictions on rural marketing to reinforce the intended incentive effects. Given the ad hoc way in which reforms have been introduced, however, there are also significant contradictions among them. The rapid spread of profit retention, for instance, reduced state revenue income at a time of growing concern over budget deficits and far outpaced other provisions of enterprise reform (such as labor-recruitment practices).

Inconsistencies are found not only among the economic reforms themselves, but—even more glaringly—also between these economic programs and other social and political innovations. A striking characteristic of Deng Xiaoping's reform movement, particularly when compared with reform movements in other socialist states, is the comprehensiveness of the Chinese effort. Virtually no facet of Chinese life has escaped untouched by the flood of new policy directives. While changes in agriculture and industry have attracted the greatest notice by outside observers, reforms in the Party, government, legal system, education, culture, health care, and family planning—to name but a sample—have also been of monumental importance.

One of the most important areas of potential conflict with the economic reforms is in the political arena. To carry out the major shifts in economic policy, committed Party leadership is essential. Yet many of the economic innovations undercut the authority and allegiance of Party members. As Richard Latham's chapter demonstrates, the reforms in agricultural responsibility tend to undermine the prestige and political advantages of grassroots Party leadership, and leave rural cadres at an economic disadvantage as well. Personnel reductions and uncertainties surrounding cadre remuneration have fueled a barrage of criticism by local leaders on grounds of ideology, social welfare implications, and practical management considerations. The result has been a serious decline in cadre commitment.

Weakened cadre allegiance to state initiatives, Elizabeth Perry proposes, is a key ingredient in the recently reported upsurge in collective violence and "feudal superstition" in the countryside. As local leaders come to identify more with their kinsmen and neighbors than with central directives, rural China has witnessed an unleashing of traditional patterns of communal behavior. The resurgence of such phenomena as ancestor worship, village temple construction, and inter-lineage feuds has taken place under the aegis—and often the active leadership—of rural cadres.

In the industrial sector as well, economic reforms have created difficulties for some Party members. Attempts to streamline factory management and upgrade the level of technical competence, for example, threaten many of the older, less qualified cadres. Susan Shirk argues that the recent industrial reforms have generated a considerable amount of political conflict—and not only from local cadres. By Shirk's account, the losers in the industrial reforms have been a sizable and influential group which she terms the "Communist coalition," that is, heavy industry, inland provinces, central planning agencies, and industrial ministries.

The opposition of unsympathetic Party members has certainly

not gone unnoticed by top-level leaders, who are attempting to mitigate resistance by a reform of the Communist Party itself. Party reform has included a number of different steps. One highly publicized approach, intended to effect a changing of the guard by encouraging elderly members to step aside, was the implementation of a retirement program. Generous pensions, complete with bonuses for lengthy service and continued access to certain perquisites of office, were offered as an inducement to retire. This attempt proved less than successful, however. In a society where lifetime tenure has been the norm for politicians, even a generous retirement package is not attractive enough to elicit voluntary compliance.[14] Moreover, Deng himself was responsible for the rehabilitation of some 2.9 million cadres who had been purged over the previous two decades. To expect these recent returnees to bow out immediately was hardly realistic. It quickly became apparent that structural change would be necessary if the Party was to be altered fundamentally.

At the Third Plenum in December 1978, the Party announced the founding of a Central Discipline Inspection Commission to oversee the establishment of inspection commissions that would monitor the activities of Party members at lower levels. The Fifth Plenum in 1980 saw the adoption of further structural changes: The Party Secretariat, responsible for guiding central policy, was reinstated under the leadership of Deng's protégé, General Secretary Hu Yaobang. Party schools, charged with training both old and new members, reopened. These developments were legitimized by the Twelfth Party Congress in the autumn of 1982, at which time Hu Yaobang announced that still more hurdles were being erected for Party members who wished to retain their status. Beginning the following year, all Party members would be required to re-register and only those who met certain qualifications would be permitted to remain.

Slow to get started, this most recent Party rectification drive is ostensibly directed against three types of offenders: (1) people who rose to prominence during the Cultural Revolution by "following the Gang of Four in rebellion," (2) people who are "seriously factionalist" in their thinking, and (3) people who indulged in "beating, smashing, and looting" during the Cultural Revolution years. Since approximately half the Party's 39 million members joined during the decade of the Cultural Revolution, there is much cause for anxiety among the Party ranks. Although it is still too early to know just how extensive the current rectification may prove, a number of arrests and expulsions from the Party have been highly publicized. In their effort to rebuild the CCP, Party leaders are also following a strategy employed by their predecessors; they are attempting to recruit new loyal followers. A series of *People's Daily* editorials in the spring of 1984 has advocated building a "third echelon" of Party members—young, technically competent individuals who will faithfully implement the new reforms. Intellectuals and newly rich peasants are among the groups targeted for recruitment. The expectation is that such people will find it in their interest actively to promote reform programs.

Whether this attempt to remold the Party will succeed is difficult to foresee. A different sort of Party may well be required if the economic reforms are to be implemented, but the very effort to rectify the CCP may generate so much resistance as to endanger the larger reform movement. Recent Chinese press reports make clear that the restoration of criticism–self-criticism sessions, for example, has met with much grumbling and subterfuge by seasoned Party members.

The political innovations now underway in China may, in fact, pose just as great a threat to the authority of local Party members as reforms in the economy. The campaign to separate Party and government work—which includes the call to reorganize people's communes into *xiang* governments—is one important example. Over the years, government units came to operate as little more than rubber stamps for Party initiatives. Most government posts of any importance were occupied by ranking Party members for whom the state position was largely ceremonial. But now, if we are to believe Chinese pronouncements, all this must change. Government units, as the administrative arm of the state, are being charged with a much more significant role in the political process. Overlap between Party and government personnel (for example, provincial first Party secretaries acting concurrently as provincial governors) is being reduced in hopes of better balancing the responsibilities of Party and government. In the countryside, the drive to establish township governments and to separate government administration from commune management is further evidence of an effort to beef up government authority and to separate administrative and economic decisions from ideological concerns. Reforms intended to democratize the government—such as county-level elections and greater freedom of debate at the National People's Congress—also suggest a more important role for government bodies.

To carry out its new responsibilities more effectively, the government bureaucracy has been substantially reorganized. Ministries have been merged and the number of ministers, vice-ministers, and bureau chiefs sharply reduced. A responsibility system for state cadres, which aims to define clear standards for efficiency, productivity, and quality of work, is being introduced.

Institutionalization of such an ambitious program of political reform is not an easy matter. To buttress the legitimacy of the changes, a legalization drive is also underway. The most dramatic indication of the increased role being assigned to the legal system came with the trial of the Gang of Four in the fall of 1980. In contrast to previous political purges, which had been conducted either behind closed Party doors or in the heat of mass criticism campaigns, the demise of the Gang of Four was to be sanctified by legal rituals. While there is no doubt that the trial was in fact carefully orchestrated by Party leaders, the televised courtroom setting did symbolize the importance the leadership was attributing to legal procedures. Over the next few years, promulgation of a series of new national laws—ranging from a criminal code to a contract law—further attested to the growing role of the legal profession.

Training of the intellectuals needed to design reforms such as those being implemented in the legal area was a serious problem. The Cultural Revolution had witnessed the closing or reorientation of many of China's more specialized and elite educational institutions. To make up for the deficit of trained personnel would require a change in educational policy. For the outside world, the most visible indication of this change has been the sending abroad of thousands of Chinese students to obtain the advanced training unavailable at home. Since 1978, China has dispatched some 18,000 students to 54 countries to study at government expense. Another 7,000 Chinese students are now studying overseas under a variety of private funding arrangements.[15]

Domestically, China's educational system has also seen great changes in the past few years. Research institutes and key-point schools have been revived, difficult entrance examinations for admission to higher education have been restored, work experience is no longer a prerequisite for college admission, the political content of the curriculum has been substantially reduced, graduate degrees are being offered for the first time in decades, and so forth.

Although a strengthened educational system is clearly necessary if the reform effort is to succeed, there is some contradiction between the emphasis on education and reforms in other areas. With the introduction of household production responsibility systems in agriculture, for example, many peasant families apparently prefer to keep their children at home to help out with farm work. Primary-school attendance reportedly declined in some rural areas when household farming was implemented. Another aspect of the rural reforms that threatens education is the reduced state investment in welfare. Preferring to invest its limited financial resources in economic construction, the state has grown reluctant to underwrite the costs of rural education. In Hebei province, for instance, in early 1983, provincial officials turned over the running of rural primary and middle schools to communes and brigades. Profits generated by rural industries or levies on peasant families are used to underwrite the new financial burden the local units must assume.[16] While the press has hailed the change as a victory for rural education, the disparities between rich and poor areas that such a policy will amplify raise serious questions about the quality of education in less developed regions of the country.

Another policy that conflicts with recent agricultural reforms is China's much publicized birth-control campaign. Joyce Kallgren's chapter shows that the single-child-family program faces formidable resistance in the countryside—due in no small part to the fact that it runs counter to the thrust of the newly implemented production responsibility systems. As peasants revert to household farming, the desire for more laborers (especially sons), as guarantors of family income and welfare, grows apace. At a time when welfare provisions are threatened by reduced state and collective investment, children are seen as the most reliable guarantee of old-age security. With patrilocal residence and village exogamy still the rule in much of rural China, the preference for sons—who will remain in their parents' home after marriage—is very keen. Moreover, the reduced role of the collective has made it more difficult to enforce state policy. Kallgren argues that the single-child program differs from earlier, successful Chinese social reforms inasmuch as it is difficult (1) to identify a small group of deviants who can be publically sanctioned, and (2) to allocate sufficient resources to elicit compliance.

As these various initiatives suggest, China is currently in the midst of an extraordinary reform movement. Bold innovations in agriculture and industry have been matched (albeit sometimes with contradictory effects) by comprehensive reforms in virtually all other areas of life: Party, government, law, education, family planning, and so forth. Will the endeavor succeed? Samuel Huntington's prognosis on comprehensive reform efforts in developing societies is not encouraging:

> The reformer who attempts to do everything all at once ends up accomplishing little or nothing... Here then is the reason why comprehensive reform, in the sense of a dramatic and rapid "revolution from above," never succeeds. It mobilizes into politics the wrong groups at the wrong time on the wrong issues.[17]

By Huntington's account, radical comprehensive reform is bound to fail for essentially political reasons. While the political dynamics are somewhat different in socialist systems, there are forces opposed to reforms which can be galvanized into action to defeat the innovators. Will this be the fate of China's current experiment as well? Let us turn now to some of the implications of the Chinese reform effort in an attempt to assess the possibility that the post-Mao reforms may escape such a dismal outcome.

IMPLICATIONS OF THE REFORMS

Nearly all developing societies attempt to achieve at least three goals in the course of modernization: growth, equality, and control. Growth involves an increase in national income, equality implies a fair distribution of the new wealth, and control refers to the state's capacity to maintain order in the face of rapid socioeconomic change. While these three goals are in theory complementary (increased resources should enhance the government's ability to realize redistribution and promote public order), in practice they often conflict. Deng Xiaoping and his associates have assigned top priority to rapid growth, but whether their reforms survive will depend in large part on their success in balancing economic growth with the potentially contradictory objectives of equality and control. Unless the Chinese economy evidences high growth rates in both agriculture and industry, advocates of greater equality and control—values dear to the hearts of many Maoist cadres—may well gain the upper hand.

In the agricultural sector, although the achievements to date are impressive, the central role played by price adjustments argues for caution in assessing future developments. Higher prices have indeed worked to stimulate production, but these are one-time gains that cannot be sustained in view of the government's continuing budgetary problems. To insulate consumers from higher food prices, the government chose to absorb the rising cost of procurement while holding urban retail food prices stable. The resulting gap between procurement and retail prices in staple food sales necessitated subsidies and losses that totaled 25 billion yuan in 1981, one-quarter of the combined revenues of all levels of government.[18] It was clear that the cost of price reform and quota reduction had severely strained the government purse strings. The central leadership has announced that further price hikes will not be forthcoming, and in 1982 took steps to ensure that the portion of procurement at base-quota prices is stabilized. Thus, continued growth in rural income will have to come from other sources.

Investment in maintaining and improving infrastructure is an obvious requirement for agricultural growth. Largely because of budgetary problems, the announcement made in 1978 to raise state investment in agriculture has not been realized. In fact, agriculture's share has declined from about 10 percent in the 1970s to only 6.1 percent of total state investment in 1982.[19] With rising rural incomes, the government is apparently expecting household private investment to pick up the slack left by state and collective funding. Whether households will make the requisite investment, however, will depend on the land-tenure system that develops, as well as public trust that the present system of household farming will be retained. In addition, the pattern of demand for industrial inputs, such as chemical fertilizer, farm machinery, and tools, changed with the reorganization of agriculture. Even with government directives allowing private purchase and ownership of these inputs, in many areas the supply and distribution have not kept up with demand. In addition to reforming the rural commercial network to better serve this new demand, further investment in productive capacity must be forthcoming to sustain continuing growth in agricultural output.

While the overall standard of living in the countryside has definitely improved during the past several years, it also seems likely that regional disparities have been exacerbated. Most of the increase in rural incomes has come from price rises and a shift from quota to above-quota sales—Lee Travers estimates that 44 percent of the increase in rural collective income during 1978–1981 came from the upward adjustment of base procurement prices, and another 27 percent came from over-quota sales—advantages that accrue disproportionately to richer teams with a surplus to sell. Similarly, teams located in more fertile and/or more urbanized areas are in a position to gain more from the increased market and off-farm employment opportunities.

As the chapter by Louis Putterman suggests, the production responsibility systems themselves may sometimes constitute a source of inequity and inefficiency. Under the new systems, land distribution tends to be on an egalitarian basis—by household, laborer, or number of persons in the household. To the extent that land is not cultivated equally intensively across households, however, inefficiency in land use occurs. As some households specialize "out of agriculture," the leadership seems to encourage the renting of land to more agriculture-oriented households. This should improve efficiency of land use, but can be expected to generate inequalities as land rents are bid upward and collected by households rather than collectives. Although collective ownership of land is formally retained, Putterman warns that favored long-term tenure may shade into de facto private ownership in future.

A reduced role for the collective poses yet another problem for Chinese policy-makers, namely, that of enforcing fulfillment of state plans. Press reports make clear, for example, that a good part of the dramatic shift from quota to above-quota grain procurement was unintended. During 1979–1981, when fulfillment of the procurement target as base prices fell from 90 percent to 80 percent (of a target that had been reduced from 70 to 60.8 billion *jin*), the state was forced to fill the gap with purchases at higher, above-quota prices.[20]

Because of problems such as these, Kathleen Hartford anticipates some modification in the future implementation of agricultural reforms. Pointing to concerns over national food supply, demographic control, and welfare provisions, Hartford argues that some elements of the old Maoist model may well be revived. Specifically, she foresees a possible return to unified distribution of rural income.

Problems of control are not confined to the agricultural sector. In industry, decentralization has gone beyond the government's ability to coordinate economic activities. In the absence of price and tax reform, the market is incapable of sending the right signals to economic agents. The burden remains with the administrative apparatus to intervene and guide production units in making optimal decisions; in the post-reform period, the problem of coordination is multiplied by the devolution of decision-making to local governments and organizations. The weakest links are the cadres staffing local planning committees, finance departments, and banks, many of whom have been only recently recruited. In many cases, they are poorly trained and quite inexperienced in managing large amounts of funds and materials. In the state sector, this weakness has allowed enterprises to have the upper hand in profit contract bargaining, resulting in enterprises being "responsible for profits but not losses."

Why did the reforms evolve in this manner? Naughton argues that planners had intended to implement decentralization along economic lines, by devolving decision-making authority to enterprises, rather than administrative units. Yet the reverse has happened. In his view, it is because the reforms transferred too many financial resources to enterprises and localities, leaving the central government budget in a precarious state. Recurrent budgetary crises placed the government in the passive position of reacting to events rather than shaping them. Wong suggests that administrative decentralization was also encouraged by the balance of power at the end of the Cultural Revolution. Having relinquished control over substantial portions of financial and material resources, the central government exercised only limited leverage over local decisions. This constituted the major difference between the post-Great Leap Forward retrenchment in the early 1960s and the post-Mao reform of the 1980s. Whereas the government was able quickly to reassert control over the economy in the post-Leap period, it was powerless to do so in 1978–1979.

In industry as in agriculture, many of the problems have stemmed from the partial nature and the side effects of the new measures. Efforts to redress past problems in one area produced new problems elsewhere. When reforms in the material allocation and commercial systems channeled major portions of producer and consumer goods through the market, prices and taxes began to perform much more important allocative functions. Similarly, in the investment sphere, as a result of fiscal decentralization and introduction of financial incentives in enterprises, nearly two-thirds of total investment is now guided by revenue-maximizing considerations. If the interaction of economic agents is to be coordinated through the market, not only must prices and taxes be corrected to send out the right signals, but the relationship among economic agents will have to be reformed. Clear responsibilities must be set to govern the relationship between the state and localities, the state and enterprises, and so forth. At present, there are centrifugal forces that threaten the system; the growing size and much more lucrative nature of the market sector are rapidly eroding state control, making plan fulfillment more difficult and adding strain to the state budget.

In the past few years, each budgetary crisis has necessitated reopening negotiations with local governments over revenue-sharing arrangements. Having given away most of its carrots, the central government negotiates each round from a progressively weaker position. In spite of the relentless pressures, central planners remain committed to reform, although they must find new instruments to wrest control from local governments. In 1983, a number of measures were introduced to curb local government influence and reverse the trend of declining central-government revenues. Beginning in January 1983, a 10 percent surtax was levied on extra-budgetary funds; in July this was raised to 15 percent. The tax-for-profits scheme was introduced to replace other forms of profit retention in state-run enterprises. If successfully implemented, the measure should also limit local government influence. As one would expect, however, these programs have run into local resistance. It is too early to tell whether local governments will be able to hold the line or whether, as Susan Shirk predicts, a "Communist coalition" demanding greater central control will bring about a retrenchment, thereby generating yet another stage in the "cycle of reform" which characterizes economic policy-making in socialist systems.

If the momentum of reform is to be sustained, a unified and responsive state apparatus capable of enforcing its will on society is an obvious requirement. Yet it is not clear that the Chinese state has been strengthened by recent events. In the countryside, Richard Latham argues that grass-roots Party cadres have been severely demoralized by the new agricultural policies. Alienated rural cadres, Elizabeth Perry shows, have provided the leadership for much of the recently reported rural violence. Certainly grass-roots leaders still wield considerable power in their areas of jurisdiction. As Joyce Kallgren points out with regard to the current birth-control campaign, rural authorities are capable of bringing tremendous pressures to bear upon recalcitrant couples. But these rural power-holders do not always choose to exercise their authority according to central dictates; "Some local authorities," Kallgren notes, "may be considerably more permissive than expected, given national goals." In short, there is reason to question the capacity of the Chinese state to act as a unified entity in enforcing unpopular policies.

Further hampering the state's family planning drive has been the decline in rural health care. A recent report by William Hsiao concludes that "economic reforms in 1980 . . . inadvertently caused a once viable model health system to crumble."²¹ Whereas in 1979 some 80-90 percent of the rural populace was covered by an organized cooperative medical system, today only 40-50 percent enjoys such protection. The number of barefoot doctors per capita has dropped, while the financial burden of medical expenses borne by users has increased.²²

Retreats such as this must generate considerable controversy among Chinese leaders over the trade-offs between economic growth and other values which have been central to their socialist mode of development. Just as a combination of economic and political forces gave rise to China's post-Mao reforms, so the future of these programs will also be decided by developments in both the economy and the political arena. To date, the results of reform have been mixed and the prognosis is uncertain. That there have been impressive achievements in some areas is undeniable, but these gains have not been attained without considerable cost. At present, the central leadership obviously judges the price to be necessary and affordable; whether such a judgment can be sustained will determine the contours of the Chinese political economy in the years ahead.

PART ONE
Agriculture

CHAPTER ONE

Socialist Agriculture is Dead; Long Live Socialist Agriculture! Organizational Transformation in Rural China

KATHLEEN HARTFORD

In the space of only five years, China's agricultural organization has undergone changes of a magnitude matched only by the collectivization process of the 1950s. Most farmland is now contracted to individual peasant households. Private rural commerce in market fairs is flourishing. Peasants and handicraft workers may hire laborers or apprentices, collectives may sell off tractors or tools to individuals, and collectively owned enterprises may be sold to households or small groups. The commune structure is being formally scrapped in favor of a system that "separates economics and politics." Is collectivism dead in rural China?

The general response of most Western observers is a resounding—and generally approving—yes. In this chapter, however, I shall present a counterargument. In my view, the architects of reorganization in China's agricultural sector cannot completely dismantle the collective system. We are likely to see in future, not a "return to capitalism" in the Chinese countryside, but a reaffirmation of several key features of collectivism. Here, in support of that view, I shall offer an abbreviated description of the old model for China's agricultural collectivism; a more detailed survey of the elements of the emerging model; and finally a consideration of the factors likely to persuade China to retain more of the old model than the emerging model might lead us to expect.

THE OLD MODEL OF COLLECTIVISM

For the old model—what we have come to consider the Maoist model (based especially on the example of the Dazhai system)— we can distinguish seven principal elements of agricultural collectivism:[1]

(1) *Centralized planning,* primarily in material quantities rather than value terms. Each subordinate unit received allocations of scarce inputs and investment funding or subsidies, and in return was obligated to deliver to its superior unit a specified quantity of produce, as tax or procurement quotas.

(2) *Collective ownership* of the means of production, including, at production-team level, the land, machinery, draft animals, and, at brigade and commune level, the industrial capacity.

(3) *Large-scale production,* particularly within the production-team framework. Annual and seasonal plans determined resource use and production patterns for the entire unit, which then assigned labor, frequently in large groups on large plots of land.

(4) *Unified distribution.* Incomes of all team members depended upon the team's net income. Individuals' income from the team was distributed proportionally based upon their work-point earnings.

(5) *Economic, social, and political integration.* The ideal was to make of each people's commune a fully integrated unit of production, provision of social services, organization of political life and transformation of consciousness, and provision of consumption needs.

(6) *Welfare guarantees* for those unable to support themselves, or provision of services such as health care and education for all members of a unit.

(7) *Territorial cellularity*, in the form of maximum possible self-reliance within a nested hierarchy of production and exchange units. Individuals' needs were provided through the units to which they were formally assigned; they could not move to another unit without official permission.

In the Maoist view, these features of agricultural collectivism worked together to meet several developmental objectives: efficient use of land, water, capital, and labor resources; higher standards of living for all peasants; basic needs guarantees; demographic control in the form of a reduced birth rate and a curb on urban migration; and an adequate national food supply. However, post-Mao leaders concluded that the model, or certain major features of it, failed, and therefore have avidly set about dismantling much of it.

Post-Mao critiques have varied in emphasis and become more pronounced over time, but overall we may summarize a general critique as follows.[2] The planning system's rigid approach and concentration on physical output rather than cost-effectiveness, the irrational insistence on the-larger-the-better in scale of production, and territorial cellularity resulted in widespread, serious waste of resources. Large-scale production, unified distribution on too large a scale, subordination of economic considerations to political and social priorities, poor accounting and arbitrary assessments of money or goods in the "three-level system" (commune, brigade, and team) of collective ownership, and overly generous welfare guarantees all had deleterious effects on production incentives and, by extension, on the standard of living as well. While it was conceded that unified distribution, collective ownership, and welfare guarantees helped meet basic needs, and that the planning system helped manage the national food supply, over the long run the Maoist model's inefficient resource use and neglect of production incentives had so inhibited production as to reduce what was available to satisfy basic needs. Critiques have not carefully examined the Maoist model's effect on demographic control, but we may gather from the silence on this subject that, on balance, even the critics would concede territoral cellularity's contribution to this control.

The critics were not unanimous on all points, and at first had neither the power to make a frontal assault on the whole Maoist model, nor a clear and comprehensive sense of the alternative with which to replace it. Rather, starting with reforms in incentive systems, they have constructed piecemeal over the past five years a new conception of how a revivified socialist agriculture may rise from the rubble of the Maoist system.

THE NEW MODEL

Production Responsibility Systems (PRS)

Responsibility systems for agricultural production were experimentally introduced beginning in 1978, and have now spread to include nearly all production teams. The responsibility systems include a wide variety of practices, but generally all involve two common features: first, production contracts, concluded between the team and smaller units below the team; and second, differential compensation distinguishing the work contribution of each producer, thus satisfying the principle of "to each according to his work." Beyond these similarities, responsibility systems vary along several dimensions: in the size of units awarded contracts, in the chosen division of agricultural (or other) tasks, in the degree of team control over production plans and disposal of final product, in the methods used to arrive at team members' compensation, and in the duration of the contracts.

From the time of their introduction nationwide in 1978, the responsibility systems have changed with a rapidity that has apparently bewildered many Chinese nearly as much as outside observers. When the systems were first bruited, then CCP Chairman Hua Guofeng pointed to the fixed-compensation systems as the most desirable arrangement. As of early 1980, this type of contract still accounted for the majority of production teams. Very soon thereafter, however, the other types of PRS—dubbed "compensation linked to output" (*lianchan jichou*)—swiftly gained ground, and by 1983, accounted for nearly all production teams.³ An enormous array of such systems were current in 1979-1981. These fell into five basic types:⁴

(1) *Specialized contracts, compensation linked to output (zhuanye chengbao, lianchan jichou)*. In this system, the production team or brigade assigned contracts to teams, groups, households, and individuals in accordance with a comprehensive production plan for numerous and often quite sophisticated lines of production (for example, aquatic products, livestock breeding, and handicrafts as well as the normal crop production). Unified distribution was maintained for the entire planning unit, and work-point allocations for different tasks might be aimed at achieving relative equality of incomes in different lines of production.

(2) *Unified management, output linked to labor (tongyi jingying, lianchan jichou)*. Teams assigned contracts to groups or individuals under a unified production plan; draft animals and larger equipment were kept under collective management. Unified distribution through a work-point system was used; work points for groups or individuals were fixed for a particular task or plot of land, but required their meeting targets for output and expenditures. This method was distinguished from specialized contracting largely by the lower degree of diversification and specialization in the unit.

(3) *Production contracted to groups, compensation linked to output (baochan daozu, lianchan jichou)*. This method was basically similar to the preceding one, except that contracts were concluded only with task groups, which were assigned to large fields.

(4) *Production contracted to households (baochan daohu)*. Described below. Until the fall of 1980, this and the following form were officially forbidden by the Central Committee; but they were being developed experimentally, particularly in Sichuan and Anhui.

(5) *Tasks contracted to households (baogan daohu)*. Described below.

Among these types, there has been a fast-moving change in composition, from those relying more on centralized team planning, unified distribution, and larger work groups, to those awarding contracts on a long-term basis to individual households, and which curtail or altogether eliminate unified distribution.

HOUSEHOLD CONTRACTING SYSTEMS. The "contracting-production-to-the-household" PRSs retain the team's role in centralized production planning and unified distribution, but day-to-day management devolves to individual households. The team assigns plots of land to households by contract; allotments may be made on a per capita basis, or pegged to a combined index of household population and labor-force participation. Contracts generally stipulate the land and other inputs to be provided by the team, the output targets, and the work points to be awarded. (These factors are often referred to as the "three fixed"—costs, output, and compensation.) In addition, contracts generally specify bonuses for better-than-contracted performance, and penalties for substandard performance.[5]

An illustration of this system comes from a production team corresponding to a natural village in a mountainous area of Jiangxi.[6] The team numbers 13 households with a total population of 142, with only about 1.2 *mu* of agricultural land per capita. In June 1981, the team adopted a responsibility system locally referred to as "divide the land to the households" (*fentian dao hu*). Land was allocated to households by a 70 percent per capita, 30 percent labor-power formula. The fields were divided into three grades, determined by quality, degree of slope, and microclimatic conditions. Households drew lots for plots in each grade. Production quotas to be handed over to the team were set, across the board, at 500 catties of grain per *mu* of land; the household could keep for itself any over-quota amount. Households paid for chemical fertilizers allocated by the team according to land area. The team awarded work points for the grain delivered to the collective, and continued to use the pre-responsibility–system method for year-end grain and cash distributions. Households that could more than meet their quota obligations were free to plant some of their land in cash crops. The most prosperous household in the team—21 members, with 14 labor power, and about one-sixth of the team's land—did so, planting some 20 to 30 percent of its land in soybeans, hot peppers, and tobacco.

In the other type of household contracting system, "contracting tasks to the household," individual households receive contracts for fixed plots of land in return for fixed payments to the collective which are supposed to satisfy state and collective requirements (taxes, welfare funds, collective investments, procurement quotas). The household keeps all other produce for its own use or for sale. Unlike the other responsibility systems, this method features no "unified distribution" of the team's product. Peasants are basically self-employed and they "compensate" themselves directly with their own output. They are also responsible for providing their own food needs.

One example of this system is a team in Zhejiang's Ningbo Prefecture with approximately 200 members in 40 households, and a total labor power of about 110.[7] The cultivated land occupies about 170 *mu*. When the team first applied household responsibilties in 1982, households received land allocations based on a population-and-labor formula. The land was assigned by lot, a fairly easy arrangement here because all the land was much the same: fertile, irrigated, and flat. Most households received their allotments in one piece. The team gave each household a "schedule" (*fang'an*) for one year, specifying (1) the grain, cotton, and oil to be delivered to the state (as tax or in procurement quotas); (2) the grain and cash to be paid to the team and the brigade (for "preparedness" and "five-guarantees" reserves, depreciation, wages, and so forth); (3) the area to plant to each crop, with a 5-percent leeway for flexibility. These obligations, as well as the fees for tractor and irrigation services and the fertilizer allocations, were set by households' land area. The team's three buffaloes were distributed to the households, divided into three voluntary associations which drew their animals by lot. The collective pig farm was dissolved, but other enterprises at brigade and commune level were retained, with profits distributed to households based on their land allotments.

The term of household contracts has varied by localities. My interviews in 1983 revealed some localities using annually renewed contracts which reallocated land to allow for changes in households' size or situation; and some in which the duration was unstipulated but assumed to be long term. In the latter, land allocations were to remain unchanged regardless of household situations ("births or deaths, you don't change it"—*sheng si bubian*).[8] The Chinese press has recently hailed explicit long-term contracts, and even "land-use certificates" granted through county offices and good for fifteen to twenty years, as the latest fruitful development in the responsibility systems—"joyously received" by the peasants![9]

THE SUCCESSION OF RESPONSIBILITY SYSTEMS. Since its first encouragement of responsibility systems, the Central Committee has stressed that adoption of any PRS must be the outcome of peasants' "democratic discussion." Up until early 1982, while many provinces—particularly poorer ones—had begun moving toward household contracts, a number of others may have been expanding those forms that would rank higher on a collectivization continuum. Other areas apparently began abandoning responsibility systems after trying them experimentally (see Table 1). National policy pointed to a range of different PRSs depending upon the relative wealth of any given unit: wealthy areas, with higher commercialization and mechanization and greater commodity production, would most likely use some form of specialized contracting; the poorest, least commercialized areas would probably find household contracts most suitable; and middling areas would tend to find some intermediate system appropriate.[10]

In early 1982, the Central Committee seemed willing to solidify the status quo by announcing a phase of "summary, perfecting, and stability" for the responsibility systems.[11] Rather than stabilizing, however, the agricultural sector made an astounding further leap toward household contracting, with the proportion of teams using these systems rising from just around half in October 1981, to 74 percent in June 1982, and then to 93 percent by summer 1983.[12]

Are we then seeing a massive spontaneous surge of the Chinese peasantry back to family farming? Despite the center's stress on voluntarism, apparently not. Even in the early stages, resistance to PRSs and especially to household contracting was considerable. That resistance is attested to by the numerous articles defending household contracting that kept appearing in the economic journals and *People's Daily*. While these articles might reflect resistance merely from high-placed cadres, the work of Western scholars who have had some opportunities for field research indicates that the resistance comes from local cadres and peasants as well.[13]

This is not to say that *all* areas resisted the change; far from it. But initiatives from above were crucial in pushing the countryside to household contracting, and may have been even more important in the accelerated changes after late 1981. The transitions show a striking similarity across provinces. This is evident from the responses to my interview question on how the responsibility system was adopted in each informant's unit. From a team in Jiangxi, which "divided land to households" in 1981:

The Communist Party cadres had held a meeting at the commune. Then the team head returned and held a team cadre meeting. Cadres called the system "divide the land to the households" (*fen tian dao hu*). The cadres didn't propagandize the system; they just held a meeting [of team members] and said this was the way it was going to be done.[14]

And from a team in Shandong:

In 1979, the team did "dividing into small groups." A directive from above said to do so.... The principle of dividing [the land] according to population also came from above.... In early 1981, Hu Yaobang came to the prefecture to inspect conditions. He asked prefectural cadres what their difficulties were in dividing the land. The upper-level cadres wanted to divide to individuals because it would raise output. Basic-level cadres didn't want this, because management would become difficult, individual interests would emerge, and you'd get disputes over boundaries. "Dividing to the households" was done in June and July of 1981.[15]

Since these informants were basically favorable to the household contracting arrangement, one may be all the more certain that their descriptions of peasants' non-role in decision-making are fairly accurate. We can be less certain of the accuracy of their perceptions of the agencies giving orders—Communist Party, Hu Yaobang. But the significant point is that they *did* perceive the

TABLE 1 Incidence of Production Responsibility Systems

Area	Date	No. of Acctg. Units	% Implementing PRS	Type 1	Type 2	Type 3	Type 4	Type 5	Type 6	Other
Anhui	Dec. 1980						40			
	Aug. 1981	419,500					ca. 35	ca. 45		
	Dec. 1981						80			
Gansu	Sept. 1980		95.0				39			
	Dec. 1980						60			
	Aug. 1981						ca. 35	ca. 45		
	Jan. 1982		97.0							
Guizhou	Dec. 1980						50			
	Apr. 1981						ca. 90			
Hebei					46.9					
Henan	Mar. 1981				53.3					
	Aug. 1981	406,000	91.0		58.0	see rt.	23 see rt.	19 see rt.	see rt.	types 3 thru 6: 80
Hubei	?				21.4					
	1980		66.8[a]	5.6						
	1981			36.9	36.9					
Fujian	Mar. 1981		94.5							
Jiangxi	Apr. 1981		over 90.0							
Inner Mon.	Mar. 1981						50% of total			

TABLE 1 (continued)

Area	Date	No. of Acctg. Units	% Implementing PRS	Type of PRS	Type 1	Type 2	Type 3	Type 4	Type 5	Type 6	Other
Shandong	June 1981	nearly 400,000	over 95.0		30.3			25.5		20	"quotas for payment of a single reward production linked to teams": 6.3
	Dec. 1981	410,000	99.0		10.0	31.0		45	4	9	
Shanxi	Spr. 1981					25.4		14.5			
	Nov. 1981	120,000	86.0		55.0				31	13	
Liaoning	Mar. 1981	100,000	90; 56 "remuneration linked to output"		31.0						
Zhejiang	Dec. 1981	330,000	95.3; 70 "remuneration linked to output"								
Yuchang pref. (Hunan)	Apr. 1981	38,900	99.7								
Nantong pref. (Jiangsu)	early 1980		54.0[a]								
	late 1980		20.0[a]								
	Oct. 1981		90.0; 80.0[a]					50			
Nationally	June 1982		99.0; 90.0					74			

(continued)

TABLE 1 (continued)

Sources: BBC Selected Weekly Broadcasts/Far East: 6897/B11/8,12, 6902/B11/5-6, and 6919/B11/1; FBIS, *Daily Report: People's Republic of China*, 20 May 1981, p. K10, and 20 August 1981, p. R5; Joint Publications Research Service, *China: Agriculture*, nos. 159, 171, 173; Zheng Renjie, "Zhonggong shishi 'nongye shengchan zerenzhi' di tantan," *Zhonggon yanjiu* 15.7:85–86 (July 1981); Liu Xumao, "Jinyibu wanshan nongye shengchan zerenzhi," *Nongye jingji wenti*, 10:12 (1982).

Note: Type 1 – Specialized contracting, remuneration linked to output   Type 4 – Production contracted to household[b]
Type 2 – Unified management, output linked to labor   Type 5 – Tasks contracted to household
Type 3 – Production contracted to group   Type 6 – Small-lot contracts, fixed compensation

[a]Responsibility systems classified as "remuneration linked to output" only
[b]Many sources use this term to apply to both types 4 and 5

household contracting system to be required from on (often very) high.

One might conclude from the shift in policy applications during 1981–1982 that Chinese leaders are again lapsing into rigidity after an initial flirtation with experimental innovations and local flexibility. But there is much more at issue there than a tendency to swing between extremes. Developments beyond the production responsibility systems themselves had, by late 1981 to mid-1982, begun fundamentally to transform many policy-makers' conceptions of where the organizational reforms were to lead China's socialist agriculture in future—and to encourage them to give this "natural" trend of development a strong assisting nudge.

Further Organizational Changes in China's Agricultural System

As production responsibility systems evolved, imbalances in rural economic and managerial systems developed. These have been addressed through a variety of corrective measures. (Two such measures, pertaining to population control and local cadres' duties, are explored in later chapters by Kallgren and Latham, and therefore are not examined here.)

SURPLUS LABOR AND DEVELOPMENT OF THE PRIVATE SECTOR. Given China's high population-to-land ratio, efficiency gains occasioned by the responsibility systems were likely to be reflected in an overt rural labor surplus. Ex ante estimates by officials had put this surplus at one-third to one-half the rural labor force of 300 million. Studies of local experiences have confirmed the expectations.[16] The labor surplus is supposed to be absorbed within the rural sector.

Several policies are aimed at speeding this absorption. First, rural cadres were urged to begin by arranging contracts with those whose special productive skills require little or no land inputs (bee-keeping, livestock-raising, and so forth), thus reducing the numbers needing access to land for their livelihood. Second, official policy now permits independent economic activity by "private individuals" who may enter small-scale production, repair work, or commerce, using their own investment funds, retaining their earnings; they are responsible for their own costs, and not subject to state or collective plans or assessments. Third, in areas not using household contracting systems, up to 15 percent of the production team's land may be assigned to individual households as private plots. Households may, in addition, retain any wasteland they have reclaimed. All households are encouraged to develop "household sidelines" to use surplus household labor and supplement their incomes from "agricultural" (crop-growing) pursuits.[17]

Recent information suggests that these moves have brought about a significant shift in the pattern of the economic activity in the countryside. This is apparent in the growing number of "specialized and key-point households" producing a wide variety of commodities, and in the general burgeoning of commerce in the rural market fairs.

By late March 1983, over 2.1 million individuals (approximately 0.7 percent of the rural labor force) had licenses for full-time private commerce. Others, probably many more, took part in trading seasonally without licenses. Some areas showed much higher participation than this; one county in Anhui province, for example, boasted as many as 2 percent of peasant households engaged in private trade.[18] Individuals' full-time private trading has certainly been facilitated by the household contracting system. For example, in the wealthiest household of the Jiangxi team mentioned earlier, the paterfamilias designated one son who was to "do commerce" outside the village while his brothers and their wives managed the household land allotment.[19] However, even without household contracting, policy since 1979 has brought a tremendous "enlivening" of rural trade. As one former Shaanxi resident related, this enlivening has reached some stunning levels of activity.[20] One group of Shaanxi peasants received permission and four trucks from their brigade, bought up black beans in nearby areas, and trucked them all the way to Guangdong, where they fetched high prices. The informant estimated that the entrepreneurs made a profit of several thousand yuan on this one trip. He himself was constantly buttonholed by peasants asking, "You're a native of [a province in southeastern China]; what do people need there now?"

As a result of such heightened activity, the number of rural and urban farm commodity markets has grown, as has trade volume (see the chapter by Terry Sicular). Rural market trade volume by 1981 reached 28.7 billion yuan, or 38 percent of the value of state procurement of agricultural products.[21] The Central Committee decision early in 1983 to permit private individuals to purchase trucks and large tractors, and to loosen restrictions on commodities that can be traded in these markets, is likely to spell greater expansion of private commercial activity in rural China.[22]

The generalized expansion of household sidelines has figured as perhaps the strongest contributing factor in this expansion of rural commerce. The 1980 output value of rural household sidelines was 30.8 billion yuan, a 32.1 percent increase over 1978. The share of household sidelines in gross agricultural output value rose from 16 percent in 1978 to 18.9 percent in 1980. While later figures are not available, the proportion has probably risen further. As one indirect indicator, peasants' earnings from household sidelines nationally rose from 26.79 percent of per capita net income in 1978 to 38.06 percent in 1982.[23]

This quantitative expansion of economic activity has now given rise to a qualitative change as well, in the growing number of rural households earning all or most of their incomes from non-cropping pursuits, which are only sometimes included in team plans. Official policy now widely encourages this development, and, by mid-1983, it could be cited as a mark of success that the households specializing in non-cropping occupations and sidelines numbered 15.64 million, 9.4 percent of all peasant households; by the end of the year, 24.82 million specialized households (over 13 percent) were reported. Some areas showed much higher rates of specialization. Yiyang prefecture in Hunan, for example, reported early in 1983 that 17 percent of rural households were specialized or semi-specialized.[24] "Commodity-grain specialized households" have recently come to join the other specialized household producers.[25]

The term "specialized and key-point households" [*zhuanye hu he zhongdian hu*] has come into general use to describe this new form of organization. The distinction between the two types of households lies in the use of household labor. "Specialized households" are those in which all or nearly all labor is applied to specialized production, while in "key-point households," only "supplemental labor" (that is, labor by women, children, and the elderly) engages in specialized production. All specialization does not, however, stem from the same origins. Some "specialized households" have been assigned to the work and awarded contracts by the team; others have developed through intensification of what began as household sidelines. Whatever their origins, these households have been lauded by analysts who stress their high commodity rate (generally over 70 percent, sometimes as much as 90 percent, of the household's production is for sale), high labor productivity, and cost effectiveness. Because of their contribution to the overall rise in rural prosperity and the national supply of agricultural commodities, they are entitled to state-supplied "award grain" (grain supplied, with or without charge, in return for sale to the state of specified quantities of a given commodity), to team allocations of fodder land or feed-grains, and to bank loans.[26]

Chinese policy-makers and scholars have increasingly focused their attention on the development of specialized and key-point households, and some point to them as harbingers of the future.[27] However, it should be noted that the specialized households have also excited hostility among some cadres and peasants. Specialized households concentrating, as they generally do, on high-value products with a high commodity rate, tend to earn whopping incomes relative to the rural Chinese average. The vast majority of peasants now and for the foreseeable future will be required to concentrate on growing low-value grain in order to ensure local and national food supplies. The favored treatment of specialized households in gaining credit has placed others at a disadvantage; in the first half of 1983, specialized households received over 60 percent of total national agricultural loans, and 75.9 percent of collective agricultural loans.[28] Many peasants and cadres resent the new, and for them largely unattainable, wealth of specialized households, and have responded with a number of "malpractices," as one Chinese radio station reported:

(1) *Raising quota.* Quotas for fulfilling state plans are raised on the two households [that is, specialized and key-point households] to control the

increase of their income. (2) *Entertainment.* The two households are often requested to entertain the participants of the so-called on-the-spot meetings or cadre meetings. (3) *Encroachment.* Interests of the two households are often encroached upon under all sorts of pretexts. Some of the two households are compelled to stop their business because they cannot stand this. (4) *Fleecing.* The two households are often overcharged under all sorts of pretexts. Some people often come to gain extra advantages by various unfair means. (5) *Charity.* People stream in to borrow from the two households with the excuse that they are profiting from their commercial businesses or processing workshops. A few cadres even hint to them that they ought to pay tribute.[29]

AGRICULTURAL TECHNOLOGY AND INVESTMENT. Frequently, the first move in setting household contracts was for team members to divide up all assets previously managed by the team. Some critics have held household contracting responsible for seriously depleting assets, such as draught animals or machinery, which cannot be efficiently used by one household. One otherwise favorable report on household contracting in Hebei province noted:

On the road to Wuqiang, I saw three mechanical wells; two were stopped up with rubble and weeds and only one was irrigating. Cangzhou Prefectural Committee reports that in Xian county there are three brigades which originally had eight mechanical wells that . . . irrigated 1,300 *mu* of land; last year, after implementing the production responsibility system, they closed up all the mechanical wells, the commune members . . . dug 70 earthen wells, and could only irrigate 400 *mu*. [This report was made in the middle of a serious drought!] In Hejian and Xian counties, 1,352 head of collective livestock were sold. Cangzhou Prefectural Committee last year investigated fifteen butcher shops which butchered 1,875 cattle—one-third of these were useful cattle which should not have been slaughtered.[30]

Defenders of household contracting point to new methods to prevent the deterioration or disuse of agricultural machinery: contracting machines to individuals, selling the machines to individuals, or allowing individuals to form an association running a tractor station.[31] In Chuxian prefecture in Anhui, where in 1980 some 70 percent of production teams already used household contracts, peasants pooled their funds to buy "126 large- and medium-sized tractors, and 1,933 hand tractors, 126 pieces of processing machinery, 18 farm vehicles, and 25,298 draft animals." This exceeded the amount bought in any previous year.[32] It may also exceed the amounts available to other areas, since the supply of such machinery remains severely limited, and is not likely to increase sufficiently in the near future.[33]

In other cases, new forms of "technical responsibility systems" and "water-conservancy management responsibility systems" have developed to improve the management of collective resources not appropriate for household management. Mechanized tilling teams, irrigation teams, pest-control teams, and other technically specialized groups may form at team, brigade, or commune level. These are responsible for providing technical services for designated tasks or plots of land; they may be paid in work points, cash, or kind, depending upon the output of the land for which they are responsible.

Examples of new water-conservancy managements in southeastern Shanxi province give some sense of the range of variation in these new systems.[34] The waterworks vary in size, technical sophistication, and area served. At one extreme are the large waterworks systems which service areas of 10,000 *mu* or more. For these, the region used a responsibility system called "contracting at successive levels, responsibility to individuals, fixed compensation with bonuses and penalties." Gaoping county, for example, had such a system, including a large reservoir, 18 smaller catchment ponds, 2 underground channels, one 1,000-meter diversion pool, 3 large irrigation channels with a total length of 31 km., and an irrigated area of over 12,000 *mu* taking in 4 communes. A management bureau was set up for the area, while under it the communes established management stations; brigades, management sections; and teams, irrigation contract groups. Successive levels, from the top down, concluded contracts with their immediate subordinate levels, specifying each task to be performed, the targets for each, the responsibilities of each individual, and the basis for bonuses and penalties.

At the other extreme of managerial scale, smaller waterworks might be handled through a method called "specialized contracts, fixed-amount contracting of tasks." For example, one brigade, facing problems in managing an electrified pumping station which serviced a 350-*mu* area, contracted the station to two brigade members who managed it in addition to their regular farming activities. They received three *mu* of irrigated land, on a *baogan daohu* basis, to cover the operating expenses of the station. If costs exceeded the proceeds from these plots, the operators had to absorb the loss; if proceeds exceeded costs, they could keep 40 percent of the surplus. In 1980, the first year of the contract, the station operators earned 1,200 yuan on the plots, leaving 300 yuan after expenses.

NEW ECONOMIC COMBINATIONS. Many of the collective's functions that lapsed with the adoption of household responsibility systems are devolving to new types of organizations, especially in those areas which had been "economically backward." Such areas began with household contracting for subsistence production. But demand for specialized products, services, or investment outlets grew with increased incomes. Households or groups of households began to invest in agricultural machinery (which they may provide on contract for other households) or small-scale manufacturing or commercial ventures, or in specialized agricultural production on an expanded scale. These "new economic combinations" may compete with existing collectively owned ventures, cooperate with them, or provide goods and services beyond the collective's capacity.[35]

A report from Henan province reveals how such combines may be formed and function. In Xinyang county, 2,738 economic combinations with 12,046 member households (6.6 percent of the county total) were set up:

> These economic combines include amalgamated kilns, prefabricated construction, and transportation. Some are combined to develop breeding of animals, planting of crops, and service trades. These are combines formed by households in the same commune or production brigade, combines formed by relatives from different communes or production brigades, long-term combines, seasonal combines or temporary combines.... The smaller ones are formed by 3 or 5 households while the largest ones are formed by less than 20 households. The one in charge of a combine is an able person in production selected by the combined households themselves. The capital is raised by the combined households ... and their business accounts are independently audited. Some of the profits are used in developing production, and the rest is distributed according to shares and labor.[36]

In one such association, 8 households raised capital and cooperated with their production brigades to set up a cotton mill. The brigade provided a site and constructed the mill; the associated households provided labor of the mill; and the brigade took responsibility for selling the product.

Chinese analysts stress the difference between the collectivization process of the 1950s and these new economic combinations, for the new forms emphasize specialized production and marketing of commodities rather than subsistence-oriented production, and they may be fully responsible for their own profits and losses. Moreover, they seem ordinarily *not* to be organized along strict territorial lines; members may belong to different teams, brigades, or even communes.[37]

The official praise for the new organizational forms emphasizes their voluntary character:

> The elementary associations emerging within the villages are not administrative concoctions, they are not collected because of high officials' will; they are organizations that the peasants establish entirely on the basis of equality, voluntarism, and mutual benefit; they are in keeping with the internal demands of socialized large production and the development of commodity economy.[38]

Nevertheless, some press reports suggest that local authorities may try prematurely to push households into these associations, though it is unclear whether this stems from over-optimism about their readiness for larger-scale operations, or from a preference for collective operations.[39]

REVAMPING COMMUNE STRUCTURES. The revised national constitution has stipulated the gradual divorce of governmental functions from the commune economic structure.[40] Great changes have also been in motion within the commune's three-tiered economic system. It is still too soon to tell how complete these changes will be, or whether they will take one or several forms nationally. Pilot projects are now being conducted throughout China to test methods for making the organizational changes.

Three counties of Sichuan province have received widest attention for these experiments. Their disparate organizational restructurings shared two common features: the gradual disentanglement of economic and political functions, culminating in setting up separate governmental and economic bodies; and the formation of specialized bodies which performed on a contractual basis the technical or specialized functions heretofore handled by communes or brigades. These included running industrial enterprises, providing pest- and disease-control services, retail sales, and so forth.[41] The example of one commune involved in these experiments provides a concrete sense of these changes.

Xiangyang commune in Guanghan county began experimenting with new methods in September 1979.[42] By November 1980, the commune was abolished and replaced by a township administrative unit, the commune Party committee became a township Party committee, and the commune management committee was changed into a township government. Some functions of these new bodies, however, sound very like the old:

> The township Party committee has the function of leading and inspecting production and various work in the township, and is responsible for propagandizing and inspecting the implementation of party line and policies. ... The township government supervises the administrative work of the whole township, and is responsible for administrative villages' construction, finance and grain-tax collection, social order, ... and for ensuring that tasks assigned by higher levels of government are done.[43]

At the next level down, production brigades were replaced by "administrative villages," each with a village head and one secretary. A village Party branch replaced the brigade Party branch, with the branch secretary (usually concurrently an agricultural cadre or village head) in charge of "grasping agricultural production," while the assistant secretary is charged with Party work. This paralleled the situation in the township, where, although governmental posts were supposed to be elective, it was stipulated beforehand that a Party committee vice-secretary was to serve as township head. One may remain skeptical of just how much "separation" is actually going on. As one informant remarked, "I've heard about these experiments, but they won't make much difference. First it's a commune revolutionary committee, then it's commune Party branch; now it's township government. It's just a change in name but the people are always the same." This opinion is reportedly shared by many of the cadres entrusted with making the local reforms.[44]

The economic reorganization within the three-tiered system may have more significance over the long term. In Xiangyang commune, economic reorganization began with transforming commune- and brigade-level industries, commerce and technical services into "industrial companies," the "commercial company," and the "agricultural technical services company," overseen by an "agriculture-industry-commerce integrated company" (AIC). For the industrial companies, teams and individuals received joint-stock shares proportional to their contributions to establishing or working in the industries; profits were to be distributed according to shares. The commercial company took over the functions and staff of the commune's Supply and Marketing Cooperative, but remained subject to the county-level (state-run) Supply and Marketing Cooperative as well as to the AIC. State ownership, "routes for commercial circulation," disposal of capital and profits, and the status of employees remained unchanged. The technical services company consisted of former commune agricultural cadres and the brigade and team agricultural technicians. The company oversaw 5 specialized stations (agricultural technology, agricultural machinery, crop protection, diversification, and management) servicing the whole township, and supervised the agricultural service technicians in each administrative village. Technicians were paid wages financed by the service company's own profits and by profits paid into the AIC by the industrial companies.

At the lowest level of organization, the production teams became agricultural cooperatives, which retain most of the team's functions (including unified distribution). The cooperatives used "specialized contracting" responsibility systems to organize production and accounting, and arranged for specialized services from the technical services company on a contractual basis.

The Sichuan experiments have recently been followed by experimental separations of government and economic administration, and managerial restructuring, in provinces as disparate as Nei Monggol (Inner Mongolia), Hunan, and Guangdong. These are announced as preludes to province-wide changes based on the experimental outcomes.[45]

Above the commune level, a few novel ventures have emerged combining the productive and marketing activities of groups of state farms, communes, urban factories, and state marketing agencies. These "agro-industrial-commercial enterprises" may specialize in only one product, or may link an entire range of products in a network overarching the original units, and sometimes by-passing the usual commercial channels. For example, the "Changjiang Integrated Company" in Sichuan, beginning with the merger of 26 state farms, later "widened its scope of operation" to communes, and

> now has 23 production centres, four branch companies at the county level, two specialized branch companies, more than 50 plants and workshops for processing farm and sideline products and over 50 stores.... [T]he company has been co-operating with some factories in the vicinity along five specialized lines, and has established economic contracts with over 100 departments in a dozen or so provinces and municipalities.[46]

More recently, however, and especially since the center's decision to stress "planning as central, market as secondary," supra-commune commercial improvements have tended to stress the role of the established state commercial network, frequently with contracts that specify in advance the quantity, quality, and price of procured goods that are to pass from households to team (or brigade or commune) to the usual commercial organs.[47]

THE EMERGING NEW MODEL

The organizational formula for Chinese agriculture is still very much in flux. On the one hand, the leadership asserts that the responsibility systems—and therefore household-based production—will remain unchanged for "a certain time," that is, for an uncertain but supposedly lengthy period. On the other hand, there has been a growing emphasis on building up more cooperative or collectivized organizational forms, whether in specialized production, in small-scale manufacturing, in commercial activities, or in agricultural infrastructure investments.[48] Just how the "centralized" and "decentralized" forms will interact, and which will be paramount, is anyone's guess. Any informed guess based on past practice, however, would at least point to an enforcement from above of a definite model once the leadership has winnowed out of the welter of local initiatives something it thinks fits together as a workable, dynamic combination.

Thus far, there has been a dramatic shift in the conception of long-run organizational developments. That shift is exemplified in the prognoses of agricultural policy brain-truster Du Runsheng.[49] In 1981, Du offered a schema of responsibility systems following the early three-track conception: specialized responsibility systems for advanced areas, household contracting for the most backward, and some intermediate form for those in between. Du, like others at the time, saw household contracting as a way of going back to the point where collectivization got off the track for the backward areas; of renewing peasants' faith in the collective; and of leading them gradually into larger-scale, more collectivized, and therefore more truly socialist forms, once the damage of over-hasty collectivization had been undone.[50] By 1983, we can find Du discussing only the development of a "cooperative economy" in agriculture. Furthermore, on the issue of development toward "large-scale" production, commonly believed an essential element of a more socialist economy, Du offered an amazing observation on household-based, "small-scale" production:

As long as we have utilized modern science and technology, conducted intensive management, and realized socialization of production on the basis of a division of labor and of tasks [read: specialization], then in the same way it can be considered modern large[-scale] economy.[51]

Du is by no means alone in this view. Increasingly, the Chinese commentaries have pointed toward a new model of socialist agricultural organization which departs from the Maoist one in virtually every respect except for collective ownership of the basic means of production. While maintaining the necessity of centralized planning, the new model points to scientific and economic planning rather than mere planning in physical quantities without regard to regional comparative advantage. Most production is to be organized on a household basis, with increasing specialization and rising commodity rates providing the dynamism needed for growth. "Socialization" of production may progress as household producers themselves recognize the greater gains made possible by combining capital and coordinating specialized labor—or merely as they become more specialized and more deeply involved in the commodity economy. The self-sufficiency of territorially defined units will be jettisoned in favor of exchange relations voluntarily entered into by all concerned parties. All production and provision of essential services will be organized for maximum efficiency and high rates of return, through contractual arrangements between the relevant parties—be they state purchasing agencies and the team (perhaps renamed a "cooperative") or households or "economic combinations"; or between household producers and specialized service "companies" or individual service vendors. Collective ownership of land and planned allocation of contracts will ensure access by all peasants to some means of making a livelihood, and will guarantee the careful husbanding and development of agricultural resources.[52] (But "collective ownership" itself is to take on new meanings for means of production other than land, for those investing in voluntary "economic combinations" will do so on the basis of shares, which will make some members' say in these new collectives more equal than others'. And, in such associations, the Central Committee noted, the remunerative principle of "to each according to his work" [a "thoroughly socialist principle"] has now to be tempered by some remuneration according to share capital.) We have, in effect, a new model of agricultural socialism, knit together by contract rather than by collectivist vision.

We see, with the emergence of this new model, a determination by China's new leaders to bury the Maoist model once and for all. In the space of only four years, they have moved from tentative tinkerings to massive dismemberment of the basic features of the former collectivized system.

On the surface, it would seem that the reforms are working well. As the chapters by Terry Sicular and Lee Travers will demonstrate, agricultural output and rural incomes have both increased substantially since implementation of the rural reforms.

These increases have been explained by many Chinese commentators as due in large part to the new responsibility systems. In my view, however, much of the success cannot be traced directly to the organizational reforms. Other factors that may have played a role include a jump in producer prices, increased private marketing opportunities, a rise in fertilizer application, and a pronounced shift in cropping patterns to capture comparative advantages.[53] The point is not that reorganization has had no effect on productivity, but that the total increase in output owes much to other circumstances.

What of the difficulties ahead for the organizational reforms? Like the Maoist model, the new model's coherence in theory must be matched in practice; all parts must be in place and playing their assigned roles if the model is to work as envisioned. Here the problems are likely to arise from a paradoxical situation: that the reforms will go too far, and that they will not go far enough.[54] Four features of the model probably cannot be taken far enough, and two already give signs of going too far.

Of the reforms that will not go far enough, the first is the separation of political and economic functions. This is intended to facilitate sound economic management, so that economic decisions can be made on the basis of economic criteria rather than of the political priorities of the moment. Thus, local political cadres should not be able to interfere with the administration of economic units or use their funds at will. While some improvement in accounting methods in local economic units might come out of this reform, it is unlikely that economic management can be liberated from Party interference. Quite apart from the overlap in personnel at the local level, one need only note the irony of the CCP Central Committee (which still maintains for itself the prerogative of proclaiming the major national agricultural policies) insisting upon local Party cadres' non-interference in economic management—while maintaining that Party members are responsible for implementing Party policies. If politics becomes truly separated from economics at the grass-roots level, it will only be because politics has lapsed entirely, something the Central Committee is by no means anxious to see happen.

Second, under the new model, all management of economic

relations is to be done through contractual agreements, in which the responsibilities of all parties are spelled out, the exchange to be made is quantified, and penalties are stipulated for non-performance. This may be workable where only two parties are involved; or where the contracting parties have a choice among contractual partners, so that poor performance by one does not force the second party into defaulting on other contractual obligations. The former situation obtains only in the most backward of areas; as for the latter, very few alternative suppliers or buyers are available in the Chinese countryside.

Third, agricultural restructuring in line with comparative advantage has already encountered major obstacles. These may be due largely to the inherent contradiction between comparative advantage and planning. By early 1982, the state placed a three-year freeze on "both the level of resales of grain to rural producers of non-grain crops and the level of inter-provincial cereal transfers."[55] This decision, stemming from policy-makers' concern for ensuring national food supply and holding down food subsidy costs (often defined in terms of limiting non-urban demands for marketed grain), places severe restrictions upon the capture of regional comparative advantage. In addition, as Lardy points out,[56] the dominant conception of comparative advantage in agricultural patterns still relies upon planning from above—though now presumably in keeping with more scientific criteria—rather than on decisions made by producers at the bottom. This may distort the perceptions of comparative advantage in any given area, leading to faulty production decisions. Finally, the current policy of concentrating state investments in "commodity bases" (especially for grain) may only skew development greatly toward areas which have significant, but not overwhelming, natural advantages to begin with, without developing *any* of the comparative advantages of the areas less blessed.

Fourth, there are already apparent difficulties for establishing more "organic" commercial ties. The restraints placed on inter-provincial grain transfers are but one indication of the resistance, at all levels of the economic system, to dilution of the state marketing system's virtual monopoly control over commercial transactions. Resistance at lower levels is harder to pinpoint, given that it must of necessity remain indirect and covert. But numerous press inveighings against foot-dragging by commercial cadres are strong signals that the fundamental transformation of commercial relations will be hard to accomplish, even assuming that the central leadership is unanimous on its desirability.

Thus, the Chinese system remains considerably more rigid than the reform rhetoric would lead us to suppose—and those rigidities are often safeguarded in practice by the very policy-makers who have ostensibly approved the rhetoric. But residual rigidities are not the only stumbling block for implementing the new model. In some respects, the organizational reforms may have gone too far. I shall focus here on the two developments which I see as most problematic: first, the independence of the household producer; and, second, the polarization of incomes in the countryside.

There remain severe limitations on the autonomy of peasant producers in the present system, not least of which is the need for the ordinary household to produce most of its own food supply and its dependence on the collective (whether it be called a production team, a cooperative, or something else) to provide essential agricultural inputs. But many households are now freed of such restrictions—either by their access to grain or inputs in rural free markets or by their opportunities to move out of the rural community—and can afford to make their own decisions regardless of team plans or procurement quotas. Planners have to deal with the undesired consequences of this new autonomy: the abandonment of agricultural land while peasants pursue higher incomes outside of agriculture, the unauthorized shift out of grain into cash crops, the violation of the planned-births policy. The threats that such trends pose for ensuring national food supply and maintaining demographic control are not yet critical, but they are obvious, and of deep concern to policy-makers.

These trends beyond autonomy to what the planners consider anarchy may be encouraged as the specialized households develop. The encouragement of specialized households spells a strong possibility of extreme polarization of incomes in the countryside, not between localities, but within them. Curiously, while the specialized households' arrangements have been touted as particularly suited to those with special skills or talents, some recent information suggests that the opportunities for specialized-household status (which are allocated by the collective) may be disproportionately assigned to team and brigade cadres or former cadres, educated youths, and demobilized soldiers[57]—the very types who were accused of using their positions to benefit at the expense of other peasants in the pre-reform system. Thus, to social polarization there may be added an element of political resentment.

Specialized households are supposed to provide the inspiration for others to attempt to enrich themselves; in the words of Wan Li, they

> are the representatives of advanced productive forces in the countryside. ... So long as the enthusiasm of these peasants is protected, others who are still leading a fairly difficult life now would have something to look forward to and the goal of common prosperity will no longer be a pie in the sky.[58]

So long, however, as the prosperity of specialized households stems from preferential access to credit, inputs, permission for high-value lines of production, and status from connections with local power-holders, "pie in the sky" would seem rather an accurate description of other peasants' hopes for similar prosperity.

My interviews in 1983 revealed a number of instances in which, although the "eating out of one big pot" tendency in agriculture was under attack locally, cadres and most peasants agreed on the need to equalize opportunities for higher incomes by allocating access to higher income-earning possibilities fairly equally across households.[59] Local attempts to "milk" the specialized households indicate that peasants are taking the time-honored route of equalization of incomes in dealing with the new phenomenon.

Predicting the future of the reforms necessarily partakes of the nature of crystal-ball gazing, but, if the factors discussed above are given any weight, we must conclude that, *so long as current national development priorities remain unchanged,* the reforms are not likely to proceed according to the blueprint set by the new model. Furthermore, I would contend that the pressures upon the reform process coming from all relevant quarters—

national planners and policy-makers, local and regional cadres, and peasants themselves—will lead toward the restoration or reconfirmation of some (not all) of the features of the Maoist model. From the national leaders' perspective, concerns with the costs of food subsidies, security of national food supply, and demographic control seem decisive. For regional and local cadres, the concerns focus more on issues such as local growth prospects, the capacity to respond effectively to demands from above, and, of course, their own power and welfare. For peasants, guarantees of basic needs and prospects for continued increases in incomes and living standards may be taken as strong concerns. These concerns are likely to bring pressures to bear upon different facets of the reforms. National-level decision-makers are likely to resort to greater emphasis on the planning system, and on the territorial cellularity of commercial relations, trends which we see already emerging. Regional and especially local cadres, if they remain at all conscientious about discharging their duties, will probably attempt to retain, or create, as many resources under collective control as possible. New developments, such as the specialized households, new economic associations, and even the creation of "companies" overseeing technical services and rural industries, offer fruitful opportunities for enhancement of local cadres' interests. We may expect them also to continue to overstep the formal boundaries between economic and political functions. From both peasants and rural cadres, we may very well see pressures for the restoration of some forms of unified distribution. For example, pressures from higher levels to "safeguard the rights" of specialized households may impel the local community to introduce "unified distribution" of assessments on specialized households in order to resolve the issue to the satisfaction of upper levels, local cadres, and peasants alike.

This does not mean that the Dazhai model will be back. Certain features of the old model, especially the tendency to large-scale production organization, are not likely to be welcomed back. Nor does it mean that the restitution of features of the old model will perfect the agricultural system. A hybrid will probably emerge that avoids major shortcomings of both models, but that still will not solve China's problems of agricultural development. If Americans characteristically turn to a "technological fix" to solve problems, the Chinese leadership has tended to resort to an "organizational fix." The new model for agricultural organization, for all its distinctiveness from the Maoist approach, may be traced to the same root assumption that reorganizing human activities will free latent energies to permit natural, spontaneous development. That such an assumption runs counter to certain other basic assumptions, such as the primacy of Party leadership, is obvious. But what is more important is that no purely organizational formula will break through China's developmental impasse. And, more particularly, no amount of organization and reorganization will substitute for the improved technology, the massive investments, and the other concrete resources that Chinese agriculture so sorely needs.

CHAPTER TWO

The Restoration of the Peasant Household as Farm Production Unit in China: Some Incentive Theoretic Analysis

LOUIS PUTTERMAN

At the Third Plenum of the Eleventh Congress of the Chinese Communist Party in December 1978, less than three years after the death of Mao Zedung and the arrest of top "leftist" leaders, the policy of assigning responsibility for agricultural production to individuals, small groups, or households was formally approved as an option for implementation at local levels. (For details, see the chapter by Kathleen Hartford.) At first, adoption of production responsibility systems in agriculture was centered on poorer provinces and remote, mountainous areas. The early forms of responsibility system also preserved the principle of "unified management," in place since 1956–1957, wherein all agricultural produce (other than that from the small private plots) was collected by the team before any distribution to those contributing labor. As late as 1981–1982, the contracting of production down to the households was viewed by many Chinese and most foreign observers as an expedient to raise incentives and hence income levels in those "backward" areas in which the collective economy was not strong, and thus inapplicable to more "advanced" regions such as Jiangsu and Zhejiang provinces, Shanghai, or the suburban counties of other major urban areas. By mid-1983, however, contracting down to small units (usually households) and the combining of individual with unified management had become the rule all over China, with *baogan daohu* or *da baogan,* "contracting in a big way," its major form.

Although Western journalists have found it natural enough to report that "a pragmatic leadership is dismantling the commune system to restore private incentives to China's farmers,"[1] scholars have reason for some perplexity over these trends. True, stepped up rates of growth in agricultural output during 1978–1983 suggest that the reforms as a package have been successful in freeing entrepreneurial spirits suppressed during an era in which households feared economic success, lest they be criticized for capitalistic behavior. Yet, one also observes growth of output during the mid-1950s, when some agricultural progress was attributed to the process of combining household units into cooperative farms, ostensibly the very opposite of the present reforms; and respectable growth of both output and yields during the 1960s and 1970s, after consolidation of the three-tier commune system.[2] What could explain the successes associated with "running in reverse" the institutional transformation (also apparently successful) of a quarter-century earlier? Was "de-collectivization"[3] a necessary component of the reform package, which also included higher prices, freer marketing, and greater diversification?

From the standpoint of *incentives,* the touchstone of the reforms, it is not immediately clear why it was necessary to make households production units in order to link effectively work input with product reward. The prototypical capitalist enterprise practices unified management in the sense that workers are rewarded from the undifferentiated streams of enterprise revenue and finance, not by a share of the items they individually produce.

Matching of efforts with rewards does not necessitate the dismantling of multi-person or "team" enterprises. Nor is the capitalist enterprise unique in this regard; collective or workers' enterprises can also link work and income shares, for example, by using basic wage scales and wage-proportionate profit-sharing. If China's Cultural Revolution dampened individual material incentives by promoting egalitarian distribution, generating fears of personal enrichment, and reducing the economic autonomy of rural collectives as well as that of their member households, then why not reverse the trend within the framework of the classical commune system of 1962–1978—that is, by improving the work-metering quality of work-point distribution systems, removing the stigma from household enrichment, allowing greater income differentials, raising producer prices where appropriate, and increasing the decision-making autonomy of the collective units?

In this chapter, I shall summarize and discuss the answers given to these questions by Chinese officials and scholars in five provinces and municipalities of China during a study tour conducted by China's Ministry of Agriculture in August and September 1983. I shall also attempt to place these answers in the perspective of some Western economic theory on incentives, and to discuss the nature of the reformed system from some related perspectives.

ANALYTICAL FRAMEWORK

A common approach to the comparison of economic organization dichotomizes economies by viewing them as more or less "capitalistic," in which case they are chock full of material incentives, or "socialistic," which means lacking in such incentives and therefore relying on coercion or other mobilizing forces. In my view, this dichotomy is not helpful, particularly if we wish to focus on issues of incentives where team or cooperative production may be involved.

A more useful dichotomy with which to begin, in this case, is that between individual or independent production, on the one hand, and team or cooperative production, on the other. The comparison of these two forms is itself two dimensional: in the first place, technological; in the second, motivational. Technologically, independent producers combining into integrated production units may or may not be capable—assuming maximally efficient organization—of raising their total productivity via such combination; whether they can do this depends on the ways in which division of labor within the enterprise can increase productivity, and on engineering or technical factors such as surface-volume relationships, scale aspects of fuel efficiency, and so on. Assuming that economies of scale are technically realizable from the combination of individuals into teams, whether or not these scale economies are actually realized depends upon the degree to which "motivational frictions" introduced by team formation impede the most efficient use of labor.

The term "motivational friction" implies that independent producers' incentives are in some sense ideal, team producers' at least potentially not so. Why? Man being a social animal, why would it not be better to assume, with Marx, that "mere social contact begets in most industries an emulation and a stimulation of the animal spirits that heighten the efficiency of each individual workman"?[4] Rejection of this supposition derives from the view, perhaps un-Marxist but as popular in Beijing as in Chicago, that much labor is largely drudgery and that people undertake it only in proportion to their expected compensation in earned rewards or escaped penalties. Starting with this working postulate, one quickly reaches the conclusion that the independent producer, retaining 100 percent of the net product of his exertions, faces ideal work incentives in the sense that every ounce of exertion will be expended for which the value of the anticipated benefit exceeds the personal cost or sacrifice, and no more.

Since undertaking each effort, and only those efforts, for which benefits exceed costs, means maximizing welfare, any deviation of incentives from the structure facing the independent producer must invariably reduce welfare attainment. If the worker receives less than the incremental value produced by his ounce of exertion, he will abstain from some labor for which the actual fruits exceed his personal cost, for example. But how can reward and contribution to output be matched when the individual is part of a team whose output is the joint product of many hands?

It is this problem that underlies emphasis on the individual/team dichotomy. Indeed, what essentially distinguishes the team from the individual is that its products cannot be unambiguously divided into lots attributable to each of its individual members. Thus 20 peasant households remain independent producers, in our sense, so long as the crops standing in their fields can be unmistakably identified with the respective cultivators. Combine the households into an integrated working unit, erase the boundaries between their fields, and it becomes impossible to identify any particular ear or row of corn as having been produced by the Zhang household. How, then, can the Zhangs' share of team income be calculated so that the net output value of an extra ounce of effort is returned to them, inducing them to undertake the exertion provided its benefits exceed their cost?

The first important point here is that the problem of work incentives in teams is a general one, not the special province of socialist collective farms. Capitalist manufacturing enterprises which pay a set wage for meeting some basic standard of performance may feel the incentive friction of teams as fully as any socialist counterpart.

Second, theories of incentives in collective or worker-managed enterprises suggest that such enterprises are, in principle, capable of matching incremental reward with incremental output, for example when enterprise net revenue is distributed in proportion to respectively contributed labor inputs, and certain other conditions are met. This example corresponds to the work-point payment systems used classically in both Soviet collective farms and Chinese production teams, provided that a homogeneous denominating unit of basic labor can be specified and that work-point recording is accurate.

Of course, there is no way of achieving the effort-reward correspondence of the independent producer if the collective uses a strictly egalitarian distribution method, or any other method that disregards labor input when determining income shares. Yet, theory suggests that pure "distribution according to work" may sometimes generate excessive work incentives—in the sense that the cost of incremental exertion exceeds the real incremental output—and that some admixture of needs distribution would be desirable on such occasions. Indeed, the team itself, if democratic and autonomous, would have a built-in tendency to select the mix of these distribution principles that is optimal from the standpoint of individual incentives.[5]

HOW IT WORKED IN CHINA

The analytical points reviewed above might, when juxtaposed with a thumbnail sketch of China's agricultural development since 1949, suggest the following interpretation.

First, because agricultural output increased between 1953 and 1958 as the Chinese Communist Party led the peasants in formation of mutual-aid teams, elementary producers' cooperatives, and advanced producers' cooperatives, it may be supposed that the scale of the individual peasant farm resulting from post-liberation land reform was relatively non-economic, and that economies of scale were achieved by the amalgamation of farm units. In other words, the technical basis for gains from cooperation was present, and was not offset by motivational frictions during the early stages of collectivization.

Second, the drastic decline in output associated with making the people's communes production, accounting, and distribution units in 1958-1959 may be attributed to technical diseconomies of scale occurring in the transition from the smaller cooperative units or, more likely, to severe motivational frictions or disincentive effects, due to management and accounting limitations producing an extremely egalitarian income distribution, along with simple bad management, political passive resistance, bad weather, and other factors.

Third, the stabilizing of output with the establishment of small teams as production and accounting units by 1962, and subsequent modest but continuing growth in output until the late 1970s reforms, suggest that the team system, using work-point accounting to tie incomes to work done, was effective at solving its work motivation problem, capturing many available economies of scale, and serving as a basis for continued technical improvements in irrigation, mechanization, and so forth. The team may have done reasonably well at managing and rewarding labor, it has been speculated, because it was a small unit composed of households having longstanding relationships. On the other hand, teams did not perform as well as they might have during periods in which "leftist" policies excessively reduced income differentials, called for rewards based on political attitude or other non-work factors, reduced the autonomy of collective units, and dictated locally inappropriate agricultural practices.

Much of this evaluation would have been accepted by economists looking at Chinese agriculture in the late 1970s, before the nature of the latest reforms became more visible. Thus, at least some academic opinion would have suggested that the system could be improved by better incentives and decentralization, while maintaining the production team as the basic production unit. For, if team production was superior to individual "gardening," as the 1950s seemed to show, then what remained to be done was to perfect the system, particularly by allowing greater local specialization, raising certain producer prices, increasing input availability and choice on input use, and encouraging distribution according to work within the teams.

WHAT THE CHINESE HAVE TO SAY

Why did the Chinese Communist Party ultimately reject this view? One might now expect to hear something like this: (1) The commune system was weakened by egalitarianism, commandism, and overemphasis on grain self-sufficiency during 1966–1976. (2) To restore vitality to agriculture, the Party permitted local experimentation in the late 1970s. (3) Experiment showed the Party and peasants that production responsibility systems taking individuals, small groups, and especially households as units of production improved agricultural performance. (4) There is nothing wrong with retention of teams as units of production, provided that distribution is strictly according to work input; however, the peasants themselves are almost universally choosing smaller units, having been given the option.

Although some of these expected explanations seemed to be correct, there were subtle differences in the Chinese position articulated in mid-1983 which invite analysis. Most important among these is that the Chinese now espoused a sophisticated and comprehensive theory of incentive failure in collective agriculture, and that, in line with that theory, they played down the effects of the Cultural Revolution on agriculture—a striking fact in view of the general denunciation of that era!

What was wrong with the basic team production and work-point distribution system, according to the current Chinese account, was that work points could not be made adequately to reflect work input *under the concrete conditions of China's agriculture*, including its level of availability of managerial skills and the "level of consciousness" of the peasants. Work input is exceedingly difficult to measure in most agricultural labor. Work points based on task-completion, which could be considered the most "scientific" and consistent with the principle of distribution according to work, are difficult to *fix* scientifically because the best way and timing of performing a particular task, and the amount of labor required, will tend to vary with natural conditions, time of day, season, soil moisture, and other factors.[6] Attempts to adopt task-based work points entail the cataloging of thousands of distinct tasks, and require expenditures of supervisory and accounting effort that taxed the capabilities of the work teams. As a result, the simpler payment system known as "basic work points" was widely adopted. This system attached a certain number of points per workday to each worker, in accordance with that worker's age, sex, physical strength, special skills, and so forth (often with very few gradations, in practice). Once the individual's specific work-point value per day was decided, it was simply multiplied by the number of days worked in the season. The net income of the team divided by the total work points earned by all members determined the value of a single work point, and an individual's earnings would equal his or her numerical per diem weight times his/her recorded workdays times the work-point value.

The problem is that, unless assignments of individual-specific weightings are made carefully and revised frequently in response to changing performance, and unless workdays are well defined and monitored, such a system degenerates into what one Chinese expert called "distribution according to potential." In practice, the basic-work-points system was adopted precisely in order to *simplify* accounting. Variations in weightings remained crude and, much worse, workday recording often depended on mere presence in the field. Thus, distribution according to work was poorly implemented, and there was little incentive for individuals to put in their best effort.

While these shortcomings could be overcome in principle, by upgrading the monitoring and accounting capabilities of team personnel, the Chinese now assert that the difficulties of establishing distribution according to work under unified management in agriculture might be more or less insurmountable because of the special character of agriculture, which is a *natural* process, is not amenable to administrative regulation, and requires responsiveness to changing conditions and, sometimes, to each meter of soil. Many go further and argue that these factors give rise to a kind of law of the household as production unit, offering United States agriculture as evidence.

The inappropriateness of distribution according to potential, at least under existing levels of consciousness, and the impossibility of implementing distribution according to work inputs, as just discussed, leave what the Chinese call "distribution according to *result*" as the best option. If objective quantitative and qualitative measurement of work input cannot be carried out because of what Western economists would call the "bounded rationality" of man,[7] then letting the *result of work* directly measure the work itself would be a better approach. However, the result of the individual's or household's work cannot be separated from that of the team, so long as the latter is the primary production unit. Therefore, implementation of distribution according to result is dependent upon rearranging the production process so as to make households or individuals producing units in their own rights (while retaining other successful and ideologically fundamental aspects of the collective economy—to be discussed below).

ECONOMIES OF SCALE AND THE COLLECTIVE ECONOMY

While Chinese spokesmen wax eloquent on incentive frictions, they seem unable to address the other half of our analytical framework, economies of scale. Do technical economies of scale exist in Chinese agriculture? If motivational (or incentive) and organizational factors are set aside, are there or are there not ways in which the combination of household units into larger-scale team production units increases the productivity of land, labor, or both? Did this aspect of collectivization in the 1950s enhance productivity? Was there a hidden productivity cost to making households the production units again, beginning in the late 1970s? No direct answer to these questions is given.

The Chinese continue to assert that collectivization led to better utilization of labor and land, and that the progress of China's agriculture since liberation could not have taken place except through the "cooperative movement." Production responsibility systems in agriculture do *not* represent a break with collectivization, since land remains collective property, households undertake production within the context of the guiding state agricultural plan, and so on. On the contrary, what has been achieved is a gigantic breakthrough in the progress of the agricultural cooperative movement, for the greatest problem of cooperativization (as recognized by Lenin, according to an informant) is how to make the land the property of all the people, yet maintain the individual peasant's attitude of "cherishing" the land. The contracting of production responsibility from collective production teams to individual households had solved this problem very effectively, perhaps for the first time in history.

It is also argued that the benefits from collectivization are maintained under the current system. The great improvements in earthworks and waterworks, irrigation and drainage, and introduction of "scientific farming" techniques and inputs, including improved seeds, increased use of chemical fertilizers and pesticides, and so forth, did not require that agricultural production itself take place at the team level. In the "more advanced" farming areas visited, Jiangsu and Shanghai, it was stressed in addition that the brigades now maintained specialized teams or "companies" that undertook irrigation, pest-control work, and so on, were paid by the individual farming households under voluntary contracts, and remunerated their own members like employees in brigade-run factories (where other kinds of bonus and responsibility systems are in turn being introduced).

If the real benefits of cooperation could be achieved *without* amalgamation of the household farms, as this implies, then the obvious question becomes: Was the entire collectivization process a mistake? Since current doctrine denigrates the Great Leap Forward and Cultural Revolution but not the mid-1950s' formation of cooperatives or the three-tier commune system as established in the early 1960s, an outright positive answer to this question is impossible. If one asks instead whether China's agriculture would be more advanced today if the evolving varieties of production responsibility system had been introduced in 1962 (when Deng Xiaoping first unsuccessfully proposed to the Central Committee that households be the principal production units) rather than 1978, one still finds few Chinese in official capacities willing to speculate.

The idea that there *are* economies of scale in agricultural production—that bigger is better—has not entirely died out in China, even among well-briefed and officially approved informants. In discussions regarding economies of scale, some responses indicated sensitivity or embarrassment about the one-acre farm as China's norm. The speaker would suggest that the present situation is only temporary, that in the next phase, of which signs are already at hand, households would increasingly specialize, and that a smaller number of households would specialize *in* agriculture, farming larger acreages left by households specializing *out* of it. Formation of cooperatives, partnerships, or combines among small numbers of these specialized households was also stressed. The vision of large farms reborn seemed to assume a China with substantially reduced population, and with a quite high ratio of non-farm to farm employment, even in the rural areas.

INCENTIVES AND ECONOMIES OF SCALE: COMMENTARY

Before proceeding to discuss the new system itself, a few remarks on the interpretations of the Chinese, just reported, are pertinent. First, I return to the fact that my informants, whether provincial bureaucrats, academics, or commune officials, consistently asserted that there had been no real effect of the Cultural

Revolution or its "Learn from Dazhai" campaign upon the practice of distribution in the production teams, although the contrary assertion would have been the natural one at this time of continuing and relentless criticism, even vilification, of the Cultural Revolution. This striking assertion bolsters the Chinese position that reform within the rubric of team production would have been inadequate. The teams are said to have implemented the purest feasible form of distribution according to work both in the pre-Great Leap Forward years—1956-1958—and in the longer period—1962-1978—during which the three-tier commune system endured. Since distribution according to work was already being practiced to its feasible limit for some twenty years, reform within the framework of team production would not have sufficed.

The main comment to be made here, aside from holding this up as an example of Promethean striving for consistency in the policy line, is that these assertions are surely *wrong*. Numerous examples of anti-incentivist policies during the 1960s and 1970s can be cited to show that distribution according to work was impossible to implement for reasons extrinsic to the team nature of production per se.[8] For example, the movement from "task-based" work points to the less discriminating "basic work points" was not entirely spontaneous, as suggested above; in fact, it was a policy *of the center* that teams came under pressure to follow.[9]

With respect to economies of scale, on which the Chinese do not provide complete answers, an answer implied in much of their discussion is worth stating. First, it seemed to be admitted that what little was gained through specialization in field work was almost universally offset by incentive problems. Second, frequent references were made to "building on the basis of the achievements of the last thirty years." While asserting that the peasants would never be required to provide their labor without compensation in response to commands in the future, that much previous work of this kind has been poorly planned and thus largely wasted, and that such work would be carried out henceforth by duly remunerated special work teams, many noted that there had been significant accomplishments of mass labor works that formed part of the material basis inherited by the present system. It was implied, then, that the forced collectivization and arbitrary ordering about of labor by the Party and state between 1958–1961 and at times during the Cultural Revolution were inexcusable, yet that, all in all, the present increasing prosperity for the peasants *within* a framework of collective ownership of land, substantial state control of what is planted, and state and collective rights to shares of the product could not have come about without these prior stages.

This conclusion runs contrary to the conventional teleology of socialism: In China, Stalinism and forced collectivization may have been preconditions to establishing a household-based and partially decentralized rural order. The earlier stage facilitated a kind of "primitive accumulation" of capital and infrastructure that could later be reaped by state, collectives, and peasants alike under the new order; and it established a kind of legitimacy of state and collective rights to peasant output in the sense that the obligations paid upon contracted land, generally small fractions of output, must now appear as a relatively light and barely objectionable burden to the individual peasant household.

Finally, the very manner in which the "mystery" of de-collectivization was posed at the beginning of this paper—asking, in effect, why the undoing of an institutional change that seemed successful in the 1950s was called for and apparently succeeded in the early 1980s—will be rejected by many China scholars on the basis of evidence that has become available since the late 1970s. Using this evidence—for example, the fact that per capita grain consumption failed to return to peak mid-1950s levels until at least the late 1970s—it is possible to argue that the net impact of collectivization on agriculture was strictly negative.[10] There can hardly be much mystery in finding that Chinese leaders who opposed rapid collectivization in the 1950s, and who believe their fears to have been well founded in view of the actual performance of collectivized agriculture, permitted or even encouraged de facto de-collectivization upon regaining power in the late 1970s.

IS "CONTRACTING IN A BIG WAY" COLLECTIVE FARMING?

How seriously should we take the Chinese claims that the new farming system, in which households are production units and keep all of their output after payments to state and team, remains collective and socialist? In my view, these claims *can* be treated seriously; yet, the system as implemented in some areas may come close to approximating a private agriculture, and there are signs of official willingness to move further in that direction than had only recently been anticipated.

In what ways is the new system collective or socialist? A full answer requires as prerequisite a careful, more philosophical discussion of the meanings to be assigned to these terms; this will *not* be attempted here. Working with, it is hoped, intuitive and admittedly unspecified understandings of those terms, I would answer: First, as the Chinese emphasize, land is collective property. Team land allocated to member households for cultivation is "theirs" only on the terms of a contract with the team, which is of limited duration and requires uncompensated (apart from the right of land use and enjoyment of collective services) deliveries to the team, making the arrangement closely resemble a fixed-rent tenant-landlord relationship, with the landlord in this case being the collective (team) itself. Households cannot legally sell their contracted land to others, or pass it on to their heirs. Their rights to rent out or contract this land onward to other individuals during the period of their own contract vary from place to place, but are at least in principle always conditional on team approval.

The second important respect in which the new system differs from private agriculture is in the high degree of control the state continues to exercise over *what is produced* on the land. Endeavors to allow the localities and the individual peasant households more initiative in determining what to produce and especially how to produce it have not changed the basic fact that most agricultural production is dictated by crop-by-crop production plans. Plan targets set at national levels are still disaggregated in detail and step by step down to the team level, where they are passed on to the households via their individual contracts.[11]

Finally, although some steps in the cultivation process are carried out by the individual household, and although it holds residual rights to the product, some aspects of agricultural production remain within the purview of the "collective economy," and many non-agricultural economic and social activities are carried out in collective units. This factor is particularly significant in areas such as Jiangsu province, where water-pumping stations, irrigation and drainage, plant protection, seed breeding, and some plowing are all carried out collectively, and where 50 or even 80 percent of the value of output in the People's Communes comes from industry rather than agriculture.

Thus, it seems conceptually apt to say that the new Chinese model is still that of a socialist or collective agriculture, although one with the novel feature that collectives contract most of their land to their own members, giving them the strong production incentive of a 100-percent claim on marginal output. Of course, we should recognize at the same time that this is no longer a system of *collective production,* in the sense that the team character of the production process, discussed in the theoretical sections above, has been decisively left behind by these reforms.

Yet, there is reason to believe that there exist areas in which implementation of the reforms has come quite close to the restoration of an essentially private agriculture. This applies to some mountainous and marginal zones in which the government is offering unlimited rights of cultivation for thirty, fifty years and longer to those carrying out reclamation activities, without any obligations of delivery to state or collective, or of plan fulfillment. It also holds where land has been parceled out to households without clear contractual terms, where delivery obligations are extremely low or non-existent, and where cadres are inactive or corrupt in distributing land.

Moreover, official willingness to risk further the collective character of land tenure and permit essentially capitalist agricultural operations may be growing. There are signs of a trend toward longer tenure periods for land contracted to households,[12] and of preference for renting out of contracted land by individual households instead of redistribution by the teams. These developments suggest a policy of simulating private tenure within a collective tenure system that may shade into a reality of private tenure collective in name only. As mentioned, there has also been encouragement of specialization either into or out of farming, and the pinning of some hopes for the future on emergence of larger household and partnership farm units. Liberal attitudes toward private hiring of labor, and rights to own privately large means of production, are also noteworthy. If the profits of households commanding such inputs cannot be attributed to their own labor alone, then one must ask, in fact, whether there is not some degree of inconsistency with the claim that household contracting is a way of implementing distribution according to work.

OWNERSHIP, EFFICIENCY, AND EQUITY

Two prerequisites for efficient land utilization in an agricultural community are that land be distributed among users so that its total current productivity is maximized, and that each piece of land be worked in a manner consistent with maximizing its long-term productivity. The goal of maximizing current productivity may conflict with equity goals unless there is redistribution after production. Likewise, the goal of maximizing long-term productivity may conflict with the policy of collective land ownership unless it is possible to compensate for improvements and penalize for damages to land on which only use rights are distributed. Ambiguities regarding land ownership and other property rights in post-reform China can be related to these potential difficulties.

In principle, it should be possible to manage land as a strictly collective resource so as to maximize benefits to the community, which can be distributed according to agreed criteria. Each cultivator would be permitted use rights to additional land up to the point at which his output on the incremental parcel equaled the incremental output of the most productive alternative user. Such an outcome could be achieved by charging all users a rent per unit of land just high enough so that total user demand equals the community's supply of land.

Because of differences in labor power, health, and other characteristics, a system of this type would lead households to farm different amounts of land. It might therefore be necessary to tax and redistribute some of the income generated if the subsistence requirements of all households are to be guaranteed. Also, the competitive rent level might be so high (for example, in land-poor, labor-rich communities) as to leave rather meager earnings to even average- or better-than-average income households, unless some of the rents collected are themselves distributed back to the producers. With such a distribution of land among cultivators maximizing total output, however, it must be possible to make all community members better off under it than they would be under any alternative land allocation, provided that the product is (re-)distributed suitably.

With respect to long-term care of land, users' incentives would be appropriate if the community could enforce standards for the maintenance of terraces, irrigation ditches, and so forth, and charge penalties and compensate for improvements at rates reflecting value to future users. In the absence of such arrangements, care of land will simply be correlated with the expected duration of use of the particular plot.

The condition that households pay a scarcity-reflecting rent for the use of collectively owned land, however, is *not* being fulfilled in China, and any attempt to achieve such a situation would run up against competing goals and principles of the system. While bidding to contract collective resources such as fishponds reportedly occurs with some frequency, the principles governing the setting of state taxes and team obligations on contracted land in no way simulate the effects of an auction. State agricultural taxes seem to have remained about as they were originally set in the 1950s, except that they are now paid by the households themselves, rather than by the collectives.[13] The delivery obligations to teams are an entirely new institution, but the basic principle governing them is to set them at such a level that the team can cover its normal and continuing expenditures for public accumulation, welfare, and administration. This means that the crops or cash collected by the team should be about equal in value to the portions that were retained by the team from collective production under the old system. Indeed, teams' expenditures will in some areas have *declined* due to reduced functions.

Although output has reportedly increased in most rural areas, upward adjustments of either the tax or the obligation to the team are said to be largely ruled out, for the moment. Instead, a recurring theme is that the government wants to lighten the burden of the peasants (as a way of winning favor after years of disillusion, and of allowing the peasants to accumulate capital), that taxes have been reduced or eliminated in areas of acute poverty or resource shortage (partly to permit peasants to invest in development of their own farms), and that collective expenditures are being maintained at modest levels and cut back where possible.[14]

Since the effective "rent" on contracted land is probably well below a scarcity rent in most parts of China,[15] non-price rationing must govern land allocation. Most localities allocate land on a per capita basis, a per worker basis, or a combination of these, with further variation introduced by differences in the weights accorded to women, children, and so on, along both dimensions.[16] Insofar as the rationing system does not reflect production potential, this must result in an allocative inefficiency, for it makes sense for households to accept shares of contract land even when others could cultivate these more productively, and when they themselves plan to focus effort upon sideline or other activities. While that inefficiency is to a degree mitigated by greater emphasis on per laborer allocation of land, there is still a strong presumption that land distribution under the contracting system should serve justice first and efficiency second. In particular, it is viewed as necessary to provide households with enough land to grow their own food requirements—which means that the option of seeking maximum production, then redistributing, is generally not considered. "Land is too scarce to allow peasants to compete for it," is the way the matter was put by Sichuan officials and economists.

While teams themselves, however, are not serving as auctioneers of land, renting by individual households is permitted. As those with greater production capability "rent in" more land, payments are made not to the team, the official owner, but to the contractee. In other words, teams provide their members with assets in land on an egalitarian basis, then allow members, rather than the team, to collect some of the scarcity rents on those assets. This outcome may be understood in terms of official discomfort with "economic" methods of land allocation, willingness to give the peasants latitude in economic matters, interest in facilitating peasant capital accumulation and sector specialization, and unconcern with inequalities emerging from household-level activities.

Along similar lines, it appears that rental payments made by former team- or brigade-level tractor drivers to collective units which allow them to offer tractor services to individuals as private operators are also so low that a great deal of "economic rent" must be getting transferred to the former team drivers under the reforms. Casual empiricism suggests that quite a few, if not most, of the touted "10,000-yuan households," the new well-to-do peasants, are these formerly collective and now private drivers. What is interesting is that the authorities show no squeamishness about allowing individuals to collect economic rents that might otherwise have gone to the collectives, so great is their desire to be able to provide wealthy peasant *models* to the populace.[17]

As for the issue of land maintenance, instead of arrangements that motivate users to care for land as property of the collective, such as compensation rules, and so on, Chinese policy-makers appear to lean toward the administratively simpler solution of extending the horizon of land tenure itself. Since this particular solution can be perfected only by making tenure permanent and allowing land to be bought and sold, one sees in current Chinese thinking the tendency to emulate more and more closely capitalist property rights in agriculture. The social implications of these trends are likely to be played down so long as struggle against the evils of egalitarianism is the prevailing policy concern.

SUMMARY

I have attempted to analyze the post-1978 reforms in China's agricultural production structure at household and team levels from the standpoint of an economic theory of incentives, with special attention to Chinese explanations of the reforms, considering their internal consistency, relationship to our own incentive-theory framework, and plausibility in relation to "stylized facts" about China's agriculture.

The theory of work incentives says that team production is a superior alternative to production by independent workers or peasant households when there are technical economies of scale to be captured, and when "motivational frictions" due to the increased difficulty of tying rewards to work input are sufficiently minimized. The Chinese analysis holds that the motivational frictions in the classical production team (1962-1978) were substantial, because work-point accounting failed to reflect the enormous variability in effective and appropriate labor performed per unit of time. In the Chinese view, this failure was due to two major causes: the limits of accounting sophistication of team personnel, and the special, organic character of agriculture—significantly, the ubiquitous "leftist tendencies" do *not* enter as a cause. Chinese analysis does not appear to address separately the issue of technical economies of scale in production, except to maintain that such economies continue to be captured in irrigation, crop protection, and other activities, in spite of the redivision of fields and their contracting to households.

The change in the unit or scale of production has not been without its effect on property rights. Although formally collective ownership of land is retained, favored long-term tenure may shade into de facto private ownership. A related trend is seen in policies on capital ownership and labor hiring, which have become relatively unrestrictive. Most interestingly, while government policy shuns land allocation by scarcity rents as a *collective* practice, it appears to permit something similar to happen through arrangements between households. The collection of scarcity rents by households rather than teams has implications for capital accumulation, income distribution, and ideology.

CHAPTER THREE

Rural Marketing and Exchange in the Wake of Recent Reforms

TERRY SICULAR

Since 1978, China's agricultural performance has been impressive. After two decades of stagnation, agricultural production and rural incomes have risen rapidly. Economic crops and sideline products have shown the best record: The average annual growth of cotton production increased from 1.4 percent between 1957 and 1978 to 13.5 percent between 1978 and 1982, and of edible oilseed production from 1.1 percent to 22.7 percent over the same periods.[1] China has accomplished these increases in economic crop production while maintaining a 2.4-percent growth in grain production.[2] Rural incomes, at least in nominal terms, have also risen rapidly in recent years. Between 1978 and 1982, rural per capita income distributed by collectives grew at an average annual rate of 12.2 percent, and income from household sidelines at an average annual rate of 30.2 percent.[3]

Agriculture's improved performance follows a broad range of policy reforms affecting agricultural production planning, commercial planning, regulations on private enterprise and exchange, and the organization of farm production and distribution. Since these policy reforms occurred more or less concurrently, it is difficult to determine the extent to which any one reform or set of reforms explains recent agricultural trends. It is nonetheless useful to try to differentiate the impact of revised acreage targets, the household responsibility system, increased procurement prices, reformulated production and sales targets, and revived rural markets on rural marketing and exchange.

Rural exchange, whether transacted privately or through state commercial organs, plays an important role in the process of economic growth. By allowing division of labor and specialization according to comparative advantage and by providing monetary rewards for economic initiative and efficient management, exchange stimulates agricultural productivity. Increased agricultural productivity is essential to development for three reasons: (1) It allows accumulation of a surplus that can be invested in productive capital, (2) it releases labor for employment in non-agricultural activities, and (3) it can lead to higher rural incomes and increased per capita consumption.[4]

For many years, the Chinese government followed policies that tended to suppress rather than promote rural exchange. Private trade was strongly discouraged, leaving state commercial channels as the primary avenue of exchange. State commercial channels, moreover, were not used to any significant extent for intra-rural exchange, but, rather, to transfer surplus from the agricultural sector to the urban and industrial sectors. This transfer was accomplished by means of a quota-procurement system where agricultural products were procured at low state prices for export or for sale to the urban population and to industry.[5] These commercial policies enforced local self-sufficiency in rural areas, reduced specialization and thus efficiency, and so dampened growth in agricultural production and incomes.

Since 1978, the Chinese government has altered a number of

policies affecting rural exchange. Reforms have applied to three areas:
  (1) Quota reforms: reapportionment of quotas among regions, adaptation of the quota system to the household responsibility system, and modification of quota levels;
  (2) Price reforms: significant increases in state procurement prices for agricultural products, and adjustment of relative prices among products; and
  (3) Private exchange regulation reforms: reduction of restrictions on private exchange in rural areas.

In general, these policy reforms have permitted more exchange in rural areas at higher prices. Thus, they have contributed to recent growth in agricultural production and income.

The paragraphs below discuss in some detail the nature and possible impact of recent reforms in state commercial quotas, pricing, and regulations on private exchange. Throughout, attention is limited to policies affecting first-category agricultural products (*diyi lei wuzi*), that is, grains, edible vegetable oils, and cotton. These crops occupy 90 percent of China's sown area, and Chinese policy-makers consider them of strategic importance to the economy.

COMMERCIAL QUOTAS

In China, the state sets quantity quotas on farm units' trade with the state commercial system. Procurement quotas specify minimum (or occasionally maximum) deliveries of certain farm products by farm units to the state; sales quotas specify maximum supplies of certain consumer and producer goods to farm units from the state. By prescribing upper and lower limits on farm units' exchange, quotas indirectly influence production and income levels, especially when state procurement prices are artificially low and other trading opportunities are restricted.

Quota planning and enforcement have been applied most vigorously to first-category agricultural products—grains, oil-bearing crops, and cotton—which are subject to unified procurement and sales (*tonggou tongxiao*). For these crops the central government plans provincial procurement and sales targets. Provincial governments, on the basis of the targets received from the central government, then determine procurement and sales quotas for prefectures, prefectures for counties, counties for communes, communes for brigades, and brigades for production teams.[6] The quotas become more specific with respect to harvest season and type of grain or oil as they are passed down.

The trickle-down planning process does not foster an economically rational distribution of quotas. At each level it encourages competition among quota recipients for supplies of desired producer and consumer goods and for low procurement quotas. This competition tends to produce a distribution of supply and delivery quotas based on political rather than efficiency considerations. In particular, the planning process leads to an equal distribution of procurement-quota responsibilities and of input-supply quotas across farm units, regardless of differences in their resources and production conditions.

The quota-planning process described above has remained more or less unaltered during the recent reform period. Two specific changes, however, have taken place. First, greater efforts have been made to set quotas on the basis of economic considerations. This has led to some reapportionment of quotas in accord with regional resource endowments and comparative advantage. Such reapportionment has contributed to productivity gains, especially in the production of economic crops.[7] Second, following implementation of responsibility-system reforms, the quota system has been extended down to households. Procurement and sales quotas are now apportioned among households either directly or through production teams by means of a contract system.[8] In theory, the extension of quotas down to individual households could affect productivity negatively. All else being equal, the larger the size of the farm unit, the more room it has for internal division of labor and specialization. In the face of commercial quotas that restrict specialization through trade, larger farm units are therefore better able to maintain productivity through internal arrangement of production. The household responsibility system has considerably reduced the size of farm units. Without careful reformulation of commercial planning quotas on the basis of each household's land and labor resources, this trend toward smaller farm unit definition could reduce the extent of specialization and thus lower productivity.

Grain, oil-bearing crops, and cotton are each subject to different quota-planning policies and have each been affected differently by recent reforms. With respect to grain, the state follows a policy of procuring surplus grain to supply urban areas; rural areas are expected to be self-sufficient except for a small number of poor areas and areas specially designated for economic crop production. The grain procurement plan has, in general, consisted of two quotas, a fixed or base quota (including the agricultural tax) and an above-quota quota. Fixed grain quotas specify at each planning level a total weight of grain that level must deliver to the state. In principle, the fixed quota is set without change for a multi-year period. Before 1971, a policy of "Three Year Fix" (*yiding sannian*) permitted revisions at most once every three years; in 1971 with a new "Five Year Fix" (*yiding wunian*) policy, the interval was lengthened to five years.

At the national level, the state has recently followed a policy of not increasing and to some extent reducing the fixed grain quota. (It is possible, however, that some localities have experienced quota increases due to reapportionment at lower levels.) In December 1978, the Central Committee passed a resolution stating that the national fixed grain-procurement target would continue to remain unchanged at the 1971-1975 level for a relatively long period into the future. In 1979, the fixed grain quota and tax were reduced. Reductions were granted mostly to poor counties, areas that had experienced crop failures or income shortfalls during 1979, and areas where quota responsibilities had imposed a relatively difficult burden.[9] The reduction for 1979 amounted to 2.75 million tons trade grain equivalents, followed by further reductions of 0.58 million tons in 1980 and 3.55 million tons in 1981.[10] During these three years, then, the fixed grain quota and tax decreased by a total of 6.9 million tons trade grain, a decline of roughly 24 percent from the 1978 level.[11]

In addition to the fixed quota and tax, the state sets above-quota grain-procurement quotas. Above-quota quotas are in principle decided jointly by planners and producers in response to changing production conditions each year. This quota is usually set as a percentage of each year's surplus grain production beyond base quota, consumption, and feed-grain requirements. In practice, above-quota quota policies have varied considerably among localities and from year to year. In parts of central China, above-quota quotas are reportedly set at about 40 percent of surplus grain production after consultation with producers.[12] In Liaoning, the above-quota quotas are reportedly set at 70 percent or more of surplus grain production.[13]

There have been no recent publicized changes in above-quota quota policies, although the percentage principle is, perhaps, being observed more strictly in areas where local officials have repeatedly over-procured in the past.[14] Clearly, the continued existence of above-quota quotas alters the significance of policy changes regarding the fixed grain tax and fixed procurement quotas. With the existence of above-quota quotas, the Five Year Fix policy no longer implies an unchanging ceiling on compulsory sales to the state, but simply instructs that all procurement beyond the fixed quota, some part of which may still be compulsory, is considered above-quota. Therefore, the recent decision to lower fixed quotas and the grain tax implies less a reduction in sales responsibilities than an increase in the proportion of above-quota to quota compulsory sales. The distinction between quota and above-quota compulsory sales is nevertheless important because above-quota quota sales receive a significantly higher price than quota sales (see discussion on procurement prices below). Moreover, as the state recognizes, fixed procurement quotas provide greater incentive for production than variable quotas because farm units are allowed to retain all production increases. Thus, increasing the proportion of above-quota to quota compulsory sales has two opposing effects: On the one hand, it increases the total revenues a team receives for a given quantity of grain sales to the state and so encourages production, while, on the other hand, it reduces the proportion of increased output that is retained and so dampens incentives.

Unfortunately, national-level data on state base and above-quota grain procurement are incomplete. Complete time-series data are available only for total grain marketed and grain marketed net of resales to rural areas, both of which include compulsory and voluntary sales to the state, and, apparently, direct sales by farmers to individuals and organizations other than state procurement agencies (see Table 2). Since the mid-1950s, the quantity of grain marketed has fluctuated between 40 and 80 million tons, with an upward trend beginning in the late 1970s and early 1980s. As a percent of total grain production, marketed grain has declined over time. In the mid-1950s, marketed grain accounted for about 30 percent of total production; by the 1970s, grain marketing had declined to about 21 percent of total production.

Data on state taxes and procurement, which include voluntary and compulsory deliveries to the state but not sales to individuals and organizations other than state procurement agencies, are available for the 1950s and for 1976-1981 (see Table 2). These figures show state grain taxes and purchases fluctuating between 40 and 45 million tons in the mid-1950s and then increasing to an abnormally high level of 55 million tons during the Great Leap Forward years 1958 and 1959. In the late 1970s, and early 1980s, state grain tax and procurement ranged from 45 to 55 million tons, indicating a slight increase over the mid-1950s level. As a percentage of total marketing and production, however, the state grain tax and procurement declined.

Since the data on grain marketing and state procurement include both voluntary and compulsory sales, they overstate the level of compulsory quota deliveries. Little information on compulsory quota levels is available. As mentioned above, the base quota and tax were reduced in absolute terms between 1978 and 1981. Evidence suggests that most of the reduction in the base quotas and tax was offset by higher above-quota quotas.

In 1981, the base quota and tax reportedly accounted for 40 percent of state grain tax and procurement,[15] or about 22 million tons trade grain equivalents. Negotiated grain procurement was 8.6 million tons. Assuming that, after subtracting base quota, tax, and negotiated procurement, the remainder of state grain procured consisted of above-quota quota deliveries, above-quota quota procurement in 1981 would be approximately 24 million tons. In 1978, the base quota and tax were 6.9 million tons more than in 1981, or 29 million tons. Negotiated sales were 3.25 million tons.[16] These figures imply a 1978 above-quota

TABLE 2 Grain Marketing and Resales in Rural Areas, 1952-1982
(million tons original grain)

	State Grain Taxes and Procurement[a]	Total Grain Marketed[b]	Marketed Grain as % of Production[b]	Marketed Grain Resold to Rural (nongcun) Population[b]	State Resales Per Capita in Rural Areas[c]
1952	27.8	33.3	20.3	5.1	10.1
1953	41.5	47.5	28.4	11.6	22.8
1954	45.1	51.8	30.6	20.2	38.8
1955	43.0	50.7	27.6	14.6	27.5
1956	41.7	45.4	23.6	16.7	31.1
1957	39.8	48.0	24.6	14.2	26.0
1958	55.7	58.8	29.4	17.0	30.8
1959	55.9	67.4	39.7	19.8	36.1
1960	42.8	51.1	35.6	20.2	38.0
1961	—	40.5	27.4	14.7	27.7
1962	32.1	38.1	23.8	12.4	22.2
1963	—	44.0	25.9	15.0	26.1
1964	—	47.4	25.3	15.6	27.1
1965	—	48.7	25.0	15.1	25.4
1966	44.9	51.6	24.1	13.3	21.7
1967	—	49.4	22.7	11.6	18.5
1968	—	48.7	23.3	10.8	16.7
1969	—	46.7	22.1	12.9	19.4
1970	—	54.4	22.7	12.4	18.1
1971	—	53.0	21.2	13.2	18.7
1972	—	48.3	20.1	14.4	19.9
1973	—	56.1	21.2	15.1	20.4
1974	—	58.1	21.1	14.1	18.7
1975	—	60.9	21.4	16.9	22.1
1976	49.4	58.3	20.3	17.5	22.6
1977	47.2	56.6	20.0	19.1	24.4
1978	46.5	61.7	20.3	19.0	24.0
1979	53.9	72.0	21.7	20.3	25.7
1980	49-50	73.0	22.8	25.0	31.4
1981	54-55	78.5	24.2	30.0	37.5
1982	—	88.1	24.9	33.1	41.2

— indicates data not available.

[a]Nicholas Lardy, *Agriculture in China's Modern Economic Development* (New York, Cambridge University Press, 1983).

[b]*Zhongguo jingji nianjian*, 1983, p. 393. These figures are for the grain production year, April of the year listed through March of the following year.

[c]Column 3 figures divided by rural (*xiangcun*) population data in *Zhongguo Tongji Nianjian*, 1983, p. 103.

quota of about 14 million tons. If this remainder indeed represents compulsory above-quota quota deliveries, then, between 1978 and 1981, the above-quota quota rose by 10 million tons, more than offsetting reported base quota and tax reductions. The total level of compulsory deliveries in recent years, therefore, has apparently not been reduced, and may even have increased. The composition of compulsory deliveries by quota type and their geographical incidence may, however, have shifted. Moreover, since total grain production has risen considerably in recent years, compulsory quotas have declined as a proportion of total output.

State sales of grain to rural areas, like procurement, are subject to commercial planning quotas. For poor areas and economic-crop specialized areas designated as grain deficit, the state sells grain using fixed quotas set on the basis of predetermined per capita rations. In addition, state grain sales have at times been linked to the delivery of other agricultural products, for example, hogs, cotton, or sugar cane, to the state. Linked grain sales are called "encouragement grain sales" (*jiangshou liang*).

Figures in Table 2 indicate that the quantity of grain supplied to rural areas has been small in per capita terms. In the mid-1950s, grain resales to rural areas were about 15 million tons and, with the exception of 1958 through 1960, remained at about that level through 1974. These figures may include private as well as state resales. Since the rural population grew during these twenty years, resales per capita declined. In the 1950s, resales per capita averaged about 30 kilograms; by the 1960s, resales had fallen to less than 20 kilograms per capita, rising to 25 kilograms by 1979. These per capita levels represent a small proportion of rural grain consumption, less than 13 percent in 1957 and 1978, the only years for which per capita rural grain consumption figures are available.[17] Thus, almost 90 percent of rural grain consumed was produced self-sufficiently. Moreover, since some significant portion of grain resales went to non-agricultural rural residents and a small number of farmers in designated specialized-crop areas, the great majority of rural residents supplied an even higher proportion of their grain consumption self-sufficiently. By forcing grain self-sufficiency, the absence or near-absence of grain sales to rural residents hindered specialization according to comparative advantage, and so reduced rural incomes and productivity.

Since 1980, total grain resales per capita have increased substantially. This trend reflects recent policies permitting expansion of free trade in rural areas and also some increase in the level of state-planned grain resales. In recent years, for example, the state has increased the level and scope of encouragement grain sales. The increase in rural grain supplies, by allowing increased specialization, has undoubtedly contributed to recent rapid growth in the production of economic crops.

State commercial planning of oil-bearing crops has followed principles similar to those for grains: The state procures oilseeds and oils to supply industry and the urban population, and expects rural demand to be met locally. Commercial planning of oil-bearing crops differs from that of grain in two respects. First, oil-bearing crops are subject only to a single, fixed quota, and any additional sales are voluntary. Second, the state follows a policy of only procuring and not selling oils in rural areas.[18] These policies have apparently remained unchanged through the recent reform period.

Table 3 presents data for oilseed production and marketing. Between the 1950s and 1970s, marketed oil declined both in absolute terms and as a percent of production. In absolute terms, oil marketed fell from about 1.4 million tons in the 1950s to about 0.9 million tons during most of the 1970s; as a percentage of production, it fell from between 75 and 95 percent to less than 60 percent of output. Since these figures include voluntary marketing to the state and to non-state buyers as well as compulsory quota deliveries, it is not clear whether the historical decline in oil procurement reflects gradual reduction of both voluntary sales and compulsory deliveries or reduced voluntary sales but stable compulsory quotas. Since 1978, oil marketing has begun to increase both in absolute terms and as a percentage of production. This reversal probably reflects the easing of grain self-sufficiency requirements and higher oil prices, since the basic oil procurement quota policies do not appear to have changed.

The principle underlying cotton commercial planning policy

TABLE 3 Edible Vegetable Oil Production and Marketing, 1953–1982
(1,000 tons oil equivalents)

	Production	Marketed Oil	% of Production Marketed
1953	1,443	1,165	81
1954	1,522	1,474	97
1955	1,825	1,597	87
1956	1,847	1,384	75
1957	1,706	1,345	79
1958	1,963	1,241	63
1959	1,758	1,448	82
1960	914	776	85
1961	758	597	79
1962	790	484	61
1963	1,029	640	62
1964	1,395	944	68
1965	1,593	1,059	66
1966	1,666	1,057	63
1967	1,657	944	57
1968	1,539	896	58
1969	1,461	827	57
1970	1,617	895	55
1971	1,713	969	57
1972	1,713	902	53
1973	1,841	947	51
1974	1,884	981	52
1975	1,880	999	53
1976	1,630	828	51
1977	1,659	876	53
1978	2,065	1,154	56
1979	2,466	1,530	62
1980	2,745	1,953	71
1981	3,655	2,791	76
1982	4,285	3,100	72

Source: *Zhongguo tongji nianjian*, 1983, p. 394.

Note: These figures are for the oilseed production year, April of the year listed through May of the following year.

has differed from that for grain and oilseeds. Whereas the state sets grain and edible-oil procurement quotas at levels that leave a certain amount behind for rural consumption, livestock feed, and seed, cotton procurement quotas are designed to purchase as much of total production as possible. One objective of this policy has been to supplant rural handicraft production with a modern state textile industry. The state textile industry is considered more efficient than handicraft production and is an important source of government revenue.

Cotton procurement quotas require farm units to deliver to the state all cotton produced except 0.5 to 1 kilogram per farm member. This compulsory procurement policy has remained unchanged throughout the recent reform period. The state has at times provided extra incentives to encourage cotton deliveries, for example, awarding rights to buy chemical fertilizers, grain, and cotton cloth for cotton sales to the state. Since 1978, several reforms have occurred in these incentives. First, in 1978 the amount of chemical fertilizer awarded for cotton deliveries to the state was raised from 70 to 80 kilograms per 100 kilograms ginned cotton.[19] Second, in 1979, localities were allowed to award either a 30-percent price bonus or one kilogram of trade grain for each kilogram of cotton delivered to the state in excess of average annual deliveries during the previous three years.[20] These incentive reforms have encouraged cotton production and procurement, especially in grain deficit areas.

The level of cotton procurement relative to production has been high, showing a slight upward trend from about 80 percent of output in the 1950s to more than 90 percent in the 1970s (see Table 4). The extremely high marketing rates after 1978 probably reflect producers' responses to price and other incentive policies introduced in 1979. (Prices are discussed in more detail below.) The figures in Table 4 include both voluntary and compulsory deliveries to the state, with compulsory procurement undoubtedly accounting for a very large share. Trends in the level of compulsory as opposed to voluntary procurement would depend on the rate of increase in cotton production relative to the rate of increase in the cotton-producing population.

TABLE 4 Cotton Production, Marketing, and Retention, 1952–1982
(1,000 tons ginned cotton)

	Production	Marketed Cotton	% of Production Marketed	Cotton Retained by Producers[a]
1952	1,304	1,097	84.1	207
1953	1,175	1,012	86.1	163
1954	1,065	830	77.9	235
1955	1,518	1,282	84.5	236
1956	1,445	1,084	75.0	361
1957	1,640	1,419	86.5	221
1958	1,969	1,798	91.3	171
1959	1,709	1,473	86.2	236
1960	1,063	962	90.5	101
1961	800	651	81.4	149
1962	750	661	88.1	89
1963	1,200	1,070	89.2	130
1964	1,663	1,521	91.5	142
1965	2,098	2,021	96.3	77
1966	2,337	2,189	93.7	148
1967	2,354	2,141	91.0	213
1968	2,354	2,133	90.6	221
1969	2,079	1,859	89.4	220
1970	2,277	2,042	89.7	235
1971	2,105	1,899	90.2	206
1972	1,958	1,789	91.4	169
1973	2,562	2,379	92.9	183
1974	2,461	2,293	93.2	168
1975	2,381	2,210	92.8	171
1976	2,055	1,917	93.3	138
1977	2,049	1,927	94.0	122
1978	2,167	2,043	94.3	124
1979	2,208	2,159	97.8	49
1980	2,707	2,681	99.0	26
1981	2,968	2,910	98.0	58
1982	3,598	3,475	96.6	123

Source: *Zhongguo tongji nianjian,* 1983, p. 394. These figures are for the cotton production year, September of the year listed through August the following year.

Note: [a]Calculated as the difference between column 1 and column 2.

The design of cotton procurement quotas combined with the fact that many regions do not have soil and weather conditions appropriate to cotton cultivation means that the rural population must rely on the state to supply its demand for cotton and cotton textiles. Dependence on state supply is reinforced by the absence of private trade in cotton; at least through 1982, the state forbade private trade of cotton and cotton handicrafts. Thus, except for the small fraction of production retained, the level of cotton consumption in rural areas has been determined by the level of state-planned sales. Data on rural per capita cotton textile consumption show a decline from 17.5 feet in 1957 to 16.5 feet in 1978.[21] Over the same period, total retained cotton output (which may or may not be included in the above per capita textile consumption figures) has fluctuated widely, ranging from less than 100,000 to more than 300,000 tons, and with an apparent downward trend during the 1970s.

The general objectives and design of the state procurement quota system for first category products have changed little during recent years, although some effort has been made to reapportion procurement and sales quotas according to regional comparative advantages in crop production. These efforts have probably led to increased productivity and incomes, especially in regions suitable for economic-crop cultivation. The effect of state commercial quotas on income and production, however, depends not only on the quota structure, but also on the prices paid by the state for agricultural products and on the extent to which products can be traded privately.

STATE PROCUREMENT PRICES

Prices are an important determinant of agricultural production, income, and consumption. To maximize income, farm units produce crops that command the highest prices relative to their costs of production. Price increases for specific crops therefore can encourage production of these crops, sometimes by eliciting higher yields per unit land, and sometimes by causing expansion of cultivation onto land formerly planted in other crops. Increases in the general level of agricultural-product prices relative to non-agricultural prices can increase production by encouraging more intensive use of farm inputs, but do not necessarily cause substitution among crops. The general level of agricultural prices also has a significant effect on farm income.

Prices not only affect agricultural production and income, but also rural consumption. Consumption responds to prices in three ways. First, prices determine the profitability of agricultural production, which in turn determines rural income. Consumption increases with income. Second, prices of consumed goods relative to income determine the purchasing power of that income: the lower consumer-goods prices relative to income, the greater consumption. Third, changes in the relative prices of consumer goods can cause consumers to substitute cheaper for more expensive goods.

The Chinese government sets both procurement and sales prices for most agricultural products. Quota procurement prices of grain, edible oils, and cotton are set by the central government, and of other crops by central, provincial, or lower government levels. Since prices affect production, income, and consumption, the state could theoretically use pricing policy instead of commercial planning quotas to bring about desired levels of agricultural production and marketing. For both practical and ideological reasons, however, the government has been reluctant to rely on prices. On the practical side, farm units' responses to price changes are less predictable than their responses to changes in planned quantities. With quantity planning, the state specifies the desired production and procurement levels; with price planning, the state has to guess which prices will elicit the desired production and procurement levels.[22] Moreover, unless retail sales prices are increased accordingly, raising procurement prices incurs direct budgetary costs. Maintaining low agricultural prices, in contrast, allows the state implicitly to tax part of the agricultural surplus. From the standpoint of state finances, procurement quotas thus provide a less expensive way of raising procurement levels. On the ideological side, price planning relies on economic incentives and markets, whose roles in a socialist system have been debated heatedly within China. In general, until the post-Mao period, policy-makers believed that economic incentives and markets should play a limited and diminishing role. Thus quantity planning was favored over price planning.

Despite the historical emphasis on quantity planning, the

state has at times also used prices to stimulate production and procurement, as well as to increase rural income. In particular, the state has occasionally (1) increased the prices paid for agricultural products relative to prices received for industrial products sold in rural areas, and (2) altered relative prices paid for different agricultural products. The stated objective of the former policy has been to increase economic returns to agriculture, thus raising rural income and living standards. Available data indicate that procurement prices of agricultural products have increased relative to the retail sales prices of industrial products sold in rural areas. As shown in Table 5, between

TABLE 5 Price Indexes for Agricultural Procurement and for Industrial Retail Sales in Rural Areas, 1952-1982 (1952 = 100)

	Agricultural Product Procurement Prices[a]	Industrial Product Retail Sales Prices in Rural Areas[b]
1952	100.0	100.0
1957	120.2	102.2
1962	164.6	115.4
1965	154.5	107.9
1970	160.4	102.0
1977	172.0	100.1
1978	178.8	100.1
1979	218.3	100.2
1980	233.9	101.0
1981	247.7	102.0
1982	253.1	103.6

Source: *Zhongguo tongji nianjian,* 1983, p. 455.

Notes: [a]Includes above-quota price bonuses and negotiated prices.

[b]This index probably understates trends in industrial product prices in rural areas because it is based on a small bundle of industrial goods which does not include certain new, more expensive inputs and several other inputs, such as hand tools, whose prices have increased over time.

1952 and 1978, the procurement price index increased almost 80 percent, while the industrial product retail sales price index remained more or less constant. Although these indexes probably overstate the improvement in agriculture's terms of trade (see notes to Table 5), they indicate a reduction in the implicit tax on agriculture.

The state has also altered relative prices among agricultural products. Relative prices have changed due to quota procurement price revisions as well as implementation of various price-incentive schemes for above-quota deliveries to the state. Quota procurement prices for major grains and commercial crops appear in Table 6. Between 1952 and 1978, the state gradually increased quota prices for most agricultural products. Grain procurement prices on average rose from 133.2 yuan per ton in 1957 to 212.8 yuan in 1978, or 59 percent. Over the same period, the quota price of cotton rose 34 percent, rape-seed 76 percent, and peanuts 96 percent. (The average quota procurement price of oilseeds increased 91 percent.)[23] In addition, in 1970 above-quota grain and oilseed deliveries began to receive a 20-percent price incentive, which was raised to 30 percent in 1972.[24] Other agricultural products received no above-quota price incentives. With the inclusion of above-quota price incentives, between 1957 and 1978, average oilseed prices rose 148 percent and grains 107 percent (see Table 7).

A perusal of price and quantity data, however, reveals that, between 1957 and 1978, prices were not a major determinant of production and marketing levels. Over this period, even though quota procurement prices for oilseeds rose 91 percent, far exceeding price increases for grains or cotton, production increased only 21 percent, less than that of either grains or cotton, and marketing actually declined 14 percent. Grains, whose price increase ranked second, experienced the greatest production growth, but marketed output grew slowly. Cotton, whose price increased the least, experienced the greatest increase in marketed output, with moderate production growth (see Table 8).

Observed production and marketing trends are better explained by quota and production-planning policies than by pricing policy. Enforcement of rural grain self-sufficiency encouraged higher grain production. Production increases were necessary to feed the growing rural population. Grain marketing, however, did not increase as much as production, because most of the incremental production was retained for self-sufficient consumption.

TABLE 6 Quota Procurement Prices for Agricultural Products, 1952–1979[a]
(yuan per ton)

	1952	1957	1962	1965	1970	1975	1977	1978	1979
Grains,[b] average	119.2	133.2	180.2	184.8	216.4	217.6	217.6	212.8	257.2
Rice (paddy), average	113.4	123.6	165.0	169.4	196.2	196.2	196.2	—	237.8[c]
Wheat	163.0	178.6	234.8	221.2	268.6	268.6	268.6	272.2	327.6
Corn	94.4	111.6	150.6	151.6	181.8	181.8	180.8	176.0	214.4
Soybeans	—	156.8[c]	—	247.4[c]	296.6[c,e]	—	—	401.2[d]	461.4[d,c]
Commercial Crops									
Cotton	1,735.6	1,713.4	1,700.4	1,840.4	1,855.8	2,116.0	2,116.0	2,304.8	2,655.2
Edible vegetable oils, average	—	—	—	—	—	—	—	1,631.4[d]	2,038.8[d]
Rapeseed	218.6	318.8	454.8	454.8	454.8	560.0	560.0	560.6	714.6
Peanuts	327.0	387.2	607.8	607.8	607.8	760.0	760.0	760.0	965.8

— indicates data not available.

Source: Zhongguo nongye nianjian, 1980, pp. 380–383, except where noted otherwise.

Notes: [a]All prices are quota prices and do not include above-quota price incentives.
[b]The average price for wheat, corn, paddy, millet, sorghum, and soybeans.
[c]Source: Chinese Academy of Social Sciences Institute of Finance and Trade.
[d]Source: *Shih Chang*, No. 2, 15 October 1979, p. 1.
[e]This is a 1966 price, but grain prices reportedly did not change between 1966 and 1970.

TABLE 7 Quota and Above-Quota Procurement Price Indexes for First-Category Agricultural Products 1978 and 1982

	1978 (1957=100)		1982 (1978=100)	
	Quota	Above-Quota	Quota	Above-Quota
Grains	159	207	126	145
Oil crops	191	248	124	143
Cotton	134[a]	134[a]	120[a]	156[a]

Source: *Zhongguo tongji nianjian,* 1983, p. 464, provides the quota price indexes given in this table. Above quota price indexes are calculated by multiplying the quota price indexes by price incentive changes noted in the text.

Note: [a]The 1978 cotton price index given in *Zhongguo tongji nianjian,* 1983, p. 464, implies a 1978 price of 214.1 yuan per ton, which is different from the 230.4 yuan price given in *Zhongguo nongye nianjian,* 1980, p. 404 (see Table 6). Other sources give a price of 230 yuan (Hubei interviews, and a Joint Directive of Procurement, Sales, and Financial Departments, No. 241, 25 May 1978). I therefore assume the 1978 index in the *Zhongguo tongji nianjian* is a misprint. The price indexes here are the *Zhongguo tongji nianjian* indexes recalculated with a 1978 price of 230.5 yuan.

Oil crops, with a quota policy similar to that of grain, also experienced production growth exceeding marketed output growth. In contrast to grain and oil crops, cotton experienced faster growth in marketing than in production. This is consistent with the design of cotton quotas to minimize retention of output and maximize procurement.

During recent years, state quota and above-quota prices have been substantially reformed. Between 1978 and 1982, quota price increases for first-category products were fairly uniform—26 percent for grain, 24 percent for oilseeds, and 20 percent for cotton. Above-quota price reforms, however, varied between products. In 1979, price incentives for above-quota sales of grain and oil crops were raised from 30 percent to 50 percent. Starting in 1979, cotton deliveries exceeding the annual average delivered over the previous three years also began to receive a 30-percent price bonus. With the inclusion of these incentive reforms, between 1978 and 1981 the above-quota procurement price for cotton rose 56 percent, grain 45 percent, and oil crops 43 percent (see Table 7). With the price bonuses, then, between 1978 and 1982, the price of cotton increased relative to the price of grains and oil crops.

TABLE 8 Production and Procurement of First-Category Products
(million tons)

	Production[a]			Production Indexes		Marketed Output[b]			Marketed Output Indexes	
	1957	1978	1982	1978 (1957=100)	1982 (1978=100)	1957	1978	1982	1978 (1957=100)	1982 (1978=100)
Grain	195.05	304.77	353.43	156.3	116.0	48.0	61.7	88.1	128.5	142.8
Oil crops	1.71	2.07	4.29	121.1	207.2	1.345	1.154	3.100	85.8	268.6
Cotton	1.640	2.167	3.598	132.1	166.0	1.419	2.043	3.475	177.0	170.1

[a]Grain production figures are for original grain, and are given in *Zhongguo tongji nianjian*, 1983, p. 158. Oil crops (oil equivalents) and cotton (ginned) are from Tables 3 and 4.
[b]Procurement of grain is in trade grain equivalents and of oil crops in oil equivalents. Procurement data are taken from Tables 2, 3, and 4.

By raising the general level of procurement prices, recent price reforms have contributed to increased rural incomes (see Travers's chapter in this volume). By altering relative prices, recent price reforms should also have influenced the composition of production and marketing. Recent trends in the composition of production and marketed output, however, once again show little correlation with relative price increases. Between 1978 and 1982, oil crops, whose procurement prices increased the least, recorded the greatest increase in both production and marketed output, followed by cotton and then grain (see Table 8). This absence of correlation suggests that state quota and above-quota procurement prices were once again not the major determinant of production and procurement trends.[25] The rapidity of production growth following the reforms also rules out new capital construction or technical change as a major determinant. Rather, the recent rapid growth in oil-crop and cotton production and procurement can probably be better explained by the recent reforms in commercial quota policies and growth in private rural exchange that permitted increased specialization in economic crop production. Reformulation of production planning targets to reflect regional comparative advantage has probably also contributed to recent production and procurement trends.

PRIVATE EXCHANGE

Private exchange, since it offers an alternative to trade with the state commercial system, can mitigate the impact of state commercial quotas and low state procurement prices. In the absence of state-planned supply, opportunities to purchase subsistence foodstuffs through private channels allow greater specialization in commercial crop production. Higher rural market prices stimulate production and marketing, and contribute to rural income; however, high market prices also divert marketed output away from state procurement. Relative prices among products on rural markets, to the extent that they differ from relative prices paid by the state, can also influence the composition of production and marketed output.

State policies on private exchange have shifted over time.

Private trade of most agricultural products was permitted in the early and mid-1950s, and also in the early 1960s. During the late 1950s and during the Cultural Revolution (1966-1976), the state forbade private exchange of first-category products and strongly discouraged private trade of other commodities.[26] The suppression of private exchange occurred partly for ideological reasons and partly to guarantee the delivery of agricultural products to the state.

Available data on the volume of private trade in rural areas reflect the above shifts in policy. Rural trade data are most complete for the volume of grain exchanged. In 1951, grain sales on free markets were 10.9 million tons trade grain equivalents. This volume dwindled to 1.7 million tons in 1954, then rose to 3.0 and 4.3 million tons in 1955 and 1956. In 1957, with new regulations prohibiting private exchange of grain, reported grain sales on free markets declined to zero. Relative to state procurement, the volume of private grain trade in the 1950s was significant, but fluctuated. Private sales stood at 38 percent of state grain procurement in 1951, fell to 3 percent in 1954, and then recovered to 11 percent in 1956.[27] No data on private grain sales are available for the 1960s.

During the Cultural Revolution, overt private trade of grain and other products was negligible, although covert exchange undoubtedly persisted at reduced levels. In 1977, the state began to liberalize regulations on private trade: Private sale of all agricultural products except cotton and cotton handicrafts was once again permitted, although grain trade remained closely supervised. Initially, private transactions were limited to local exchange between producers, but recently collectives and individuals have been allowed to engage in long-distance trade. Wholesaling and resale of agricultural products by households, collectives, and individual middlemen is also permitted.[28]

These recent liberalizations are reflected in the volume of rural market transactions. By 1979, the volume of grain sold on rural markets exceeded 5 million tons (trade grain or original weight not specified), equivalent to about 10 percent of state grain procurement. In the same year, the total value of products exchanged on rural markets reached 17 billion yuan,[29] equivalent to about 30 percent of the total value of state agricultural procurement.[30] By 1982, the volume of rural-market trade reached 29 billion yuan, and the total number of rural markets had increased to 41,184, surpassing the 1965 count (see Table 9).

TABLE 9 Rural Markets: Number and Total Value of Transactions

Year	Number of Rural Markets	Total Value of Transactions (billion yuan)
1965	37,000	6.8
1979	36,767	17.1
1980	37,890	21.1
1981	39,715	25.3
1982	41,184	28.7

Source: *Zhongguo tongji nianjian*, 1983, p. 386.

Note: These figures do not include urban markets.

Increased opportunity to participate in rural exchange has undoubtedly stimulated agricultural production. By supplying producers with grain and oils, it has made possible greater specialization and so has increased cultivation of commercial crops. Moreover, since rural market prices for many products have exceeded state prices,[31] private trade has provided additional monetary incentives. Price quotations for 1980 and 1981 from a national survey of rural markets show rice prices exceeding the state above-quota price by more than 45 percent, wheat by 14 percent, and vegetable oils by more than 20 percent (see Table 10). Since the quality of products sold in rural markets may be higher than that of products sold to the state, and since, in the case of oils, the composition of market sales may include more higher-value oil varieties than that of procured oils, these nominal price differences could overstate the actual differences between market and state prices. Nevertheless, agricultural market prices appear to be higher than those paid by the state.

Although, by allowing opportunities to trade and offering higher prices, rural markets have encouraged production, they have also increased the difficulty of enforcing state delivery quotas. In order to prevent quota procurement shortfalls and

TABLE 10 National Average Rural Market Prices and State Above-Quota Prices for Rice, Wheat, and Vegetable Oils, 1980 and 1981

(yuan per ton)

	1980			1981		
	Rural Market Price[a]	State Above-Quota Price[c]	Market price ÷ Above-Quota Price	Rural Market Price[a]	State Above-Quota Price[c]	Market Price ÷ Above-Quota Price
Rice	525[b]	357	1.47	518[b]	357	1.45
Wheat	565	494	1.14	565	494	1.14
Vegetable Oils, average	3,944	3,058	1.29	3,674	3,058	1.20

Notes: [a] These prices are unweighted averages of monthly quotations from the National Survey of 206 Rural Markets published in 1981 issues of *Zhongguo caimao bao*.

[b] Rural market rice prices quoted in the source mentioned above are for husked rice. I have converted them to paddy prices using a conversion rate of 0.7 : 1.0.

[c] 1979 quota prices from Table 6 were multiplied by 1.5 to include the 50 percent above-quota price incentive. State procurement prices for these items have apparently not changed since 1979.

unauthorized price hikes by local procurement agencies trying to meet their quotas, the central government has stipulated that products can be sold on rural markets only after quota responsibilities are fulfilled.[32] In some areas the state has gone so far as to close rural grain markets during the grain procurement season until quotas are met.[33] The state has also prohibited procurement by local agencies at higher negotiated prices before quotas are fulfilled.[34]

CONCLUSION

Recent policy reforms affecting marketing and exchange of agricultural products in rural China include reduction and reapportionments of procurement quota levels, adaptation of the quota system to new responsibility system structure, a significant increase in procurement prices, an increase in the price of cotton relative to the prices of grains and oil crops, and considerable liberalization of restrictions on private trade. In the wake of these and other policy reforms, production, marketing, and rural incomes have risen substantially. Quota reapportionment and market liberalization are probably major factors behind the observed changes in the levels and composition of production and marketed output. In particular, these reforms can explain much of the recent rapid growth in the production and marketing of commercial crops like cotton and oil crops. Although quota and above-quota procurement price increases have undoubtedly contributed to higher rural incomes, they appear to be a less important factor explaining recent production and marketing trends. Responsibility system reforms, although they probably explain some of the overall growth in agricultural output, also do not explain the uneven growth among crops in recent years.

Even though recent policy changes have permitted more flexibility at the local level, commercial planning has continued to govern rural production, marketing, and consumption. The compulsory quota system, although slightly modified, has remained intact. With respect to first-category products, reduction of the basic grain quota and tax has lowered the proportion of base-quota deliveries in total procurement, but has not reduced the level of compulsory deliveries. Producers still face additional compulsory above-quota grain quotas. Compulsory oil-crop and cotton delivery quotas have also remained in place. With respect to other agricultural products, in particular, second-category items such as hogs, eggs, tobacco, tea, sugar crops, and silk, the state has also continued the system of compulsory procurement.[35] How long compulsory procurement will continue is, of course, not known. A recent government report by Premier Zhao Ziyang proposes gradual reduction in the role of state-planned procurement, and an increased role for "free" (*zi you*) exchange. The extent of this proposed change is unclear, especially as Zhao states it will occur under the condition that state procurement plans are met (*zai bao zheng wan cheng guojia shougou jihua de qianti xia*).[36]

Implementation of the responsibility system and other policies reducing the size of the farm unit have possibly intensified the impact of quotas on production, marketing, and consumption. Without reformulation of the quota-planning process, a decrease in farm unit size reduces the number of decisions made internally by the farm unit and increases the number of requirements imposed from outside. For example, decisions made by the production team were once internal to the farm unit, since the team was the farm unit. With household responsibility system reforms, team-level decisions are external to the farm unit (now the household). To the extent that team and household interests diverge, households may find directives of the team restrictive. Moreover, smaller units have less room to maneuver around planning restrictions: The impact of self-sufficiency constraints on numerous small units, some of which may be poorly endowed to produce grain, is greater than on fewer, larger units that can specialize internally.

Recent price reforms have clearly increased payments for compulsory deliveries of agricultural products and so have raised rural incomes. With the alternative of private trade at even higher prices, however, compulsory quotas are probably still considered a burden. In addition, the real value of increased rural incomes has been diminished by shortages in planned supply of manufactured goods to rural areas.[37] Higher incomes have led to increased rural demand for consumer and producer goods, but the quantities supplied have been insufficient.

Private trade has mitigated the impact of shortages in state-planned supply by providing an alternative source of supply to rural residents. Farm units are no longer forced to be self-sufficient in grain and oil, as they now can purchase these items on the market. However, restrictions on private exchange persist: Private grain trade is closely monitored, and private exchange of cotton, cotton textiles, and other manufactured items is also controlled.[38] The future volume of private trade and the extent to which rural residents can rely on free markets as a source of supply therefore remains in question.

CHAPTER FOUR

Getting Rich through Diligence: Peasant Income after the Reforms

S. LEE TRAVERS

In December 1978, participants at the Third Plenary Session of the Eleventh Central Committee of the Chinese Communist Party ratified major policy changes designed to increase peasant income substantially. Per capita peasant income had stagnated over the two decades prior to 1978 due to rapid increases in the cost of agricultural production, overall agricultural prices which rose little from 1961 to 1977, and a high rural population growth rate.[1] Measured in 1977 yuan, total peasant per capita income rose from about 102.8 yuan in 1957 to only 113 yuan in 1977, or about 0.5 percent per year.[2] Not only was rural income growth slow, it also lagged far behind that of the urban sector. In terms of purchases of consumer goods from the commercial system, in 1957 urban residents averaged 239.6 yuan per capita and rural residents 37.2 yuan, but, by 1978, urban residents were consuming 556.8 yuan per capita of these commodities and rural residents only 55 yuan per capita. The urban consumption increase of 4.1 percent per year far exceeded the 1.9 percent per year increase in rural areas.[3] The new policies were designed to address these low absolute and relative levels of rural income and consumption.

Professor Sicular's chapter in this volume provides an excellent overview of Chinese rural-development strategy. Here it will suffice to point out that, under the production, marketing, and price regimes employed between 1957 and 1977, crop yields rose too slowly to increase peasant income significantly.

In early 1978, then Party Chairman Hua Guofeng was advocating a highly collectivized, closely state-controlled agricultural-development strategy. Output gains were expected from specialization in collective production, with the emphasis on large-scale commodity production of grain and major economic crops. Collective sideline production was also to be increased, but private production and marketing continued to be restricted.[4] The set of policy changes approved in late 1978 challenged this model by relaxing constraints on private marketing. These initial policy changes sought income increases through measures that would strengthen the collective; later moves showed a willingness to promote income increases at the cost of collective strength.

The new policies are a clear success in raising rural incomes. Per capita annual peasant income increases, adjusted for inflation, averaged 11.4 percent between 1978 and 1981. By 1981, the urban/rural ratio of consumer goods purchased from the commercial system had fallen from the 1978 level of 10 to 1 to only 6 to 1.

This chapter analyzes peasant income by origin: from the collective, sideline, or other sectors of the rural economy. Table 11 shows the author's estimates of the magnitude and composition of peasant income and year-to-year increases, 1978–1981.

Peasant income from the collective sector is a combination of the peasant's share of net profits from team productive activities, indirect or direct compensation from commune and brigade enterprises or other commune and brigade labor, and other

TABLE 11 Composition of Peasant Income and Year-to-Year Increases, 1978–1981

	1978	1979	1980	1981
Net total (billion yuan, %) from:	108.24 (100.00)	127.76 (100.00)	145.75 (100.00)	165.94 (100.00)
Collective[a] (billion yuan, %)	63.05 (58.3)	71.02 (55.6)	73.85 (50.7)	80.75 (48.7)
Household sidelines[b] (billion yuan, %)	37.67 (34.8)	45.58 (35.7)	55.84 (38.3)	67.15 (40.5)
Other[c] (billion yuan, %)	7.52 (6.9)	11.16 (8.7)	16.06 (11.0)	18.04 (10.9)
Real income per capita peasant[d] (yuan)	137.5	156.8	171.5	189.9

Sources: All figures derived from official data. See S. Lee Travers, "Post-1978 Rural Economic Policy and Peasant Income in China," *China Quarterly*, no 98, for derivation details.

Notes: [a]Collective income is the sum of collective distributed income and commune and brigade enterprise direct wages, both drawn directly from official sources.

[b]Sideline income was adjusted from the sample survey figures by using official data on total national sideline output and on free-market sales, yielding rates of growth lower than the survey indicated.

[c]Other income is drawn from the sample survey, as there are no available alternative estimates by which to judge the survey figures.

[d]These numbers are derived from dividing net totals by commune year-end population and adjusting the cash-income component for inflation, with 1978=100.

direct payments (for example, for sales to the team of household manure). As can be seen from Table 11, collective income is the dominant source of income, though its importance and nature have both been changing, as will be explored below.

Peasant sideline income is derived principally from the production and marketing of produce from the peasant's private plot. Handicraft items produced in the household are another component of this income stream.

The Other income category includes wages and other income earned outside the commune, for example, wages directly paid to peasants serving as temporary workers in factories. It also includes remittances, both from within China and abroad, but it does not include loans.

COLLECTIVE INCOME

Gross income increases in collective agriculture occur in three ways: increases in production realized in kind or in sales, increases in income due to procurement price increases, and increases in income captured through increasing sales at above-quota or negotiated prices. The 1978 policy changes raised purchase prices and lowered purchase quotas, actions that substantially increased profitability of agriculture.

The base procurement price increase for 1979 averaged 20 percent for grain and 17 percent over all crops. The above-quota price bonus was raised from 30 to 50 percent for grain and edible oil, and a uniform bonus of 30 percent of the base price was established for above-quota cotton. Further adjustments in 1980 and 1981 increased the overall level of base prices by 3.5 percent and 2.4 percent, respectively.[5] Base-price increases alone were responsible for 44 percent of the rise in value of collective sales between 1979 and 1981. This increase was supplemented by income from above-quota sales, and quotas were lowered to allow more of these sales. Data on collective sector production, sales, and prices permit the computation of income increases by source: As Table 12 indicates, 43 percent of the 1979 income gain came from increased production, 44 percent was due to base price increases, and the remainder to over-quota sales. In 1980, only 8 percent of the gain was from output

TABLE 12 Estimated Sources of Growth in Collective Income, 1979–1981
(%)

	1979–1981	1979	1980	1981
Increased production	29	43	8	48
Base price increase	44	44	48	22
Increases in over-quota sales	27	13	44	30

Source: For details of methodology, and data sources see Travers, "Post-1978 Rural Economic Policy and Peasant Income in China."

increases, 48 percent was from price increases, and 44 percent from increases in over-quota sales. The better crop year of 1981 resulted in 48 percent of the gain coming from output increases, 22 percent from price increases, and 30 percent from increases in over-quota sales.[6]

At the Third Plenum, requisition quotas were fixed for a five-year period at the 1971–1975 level. A 1979 grain-tax reduction, discussed in detail below, allowed 2.35 million metric tons (mmt) of grain previously delivered at a zero price to be sold above quota or directly consumed. In 1980, a reduction of 0.58 mmt was made in the base-sales quota of some areas with predominantly minority group populations, and, in 1981, there was a reduction of 3.55 mmt in the base quotas of some of southern China's rice growing areas.[7] Thus, by 1981, a total of 6.48 mmt, or 15.7 percent of the annual commune grain-tax and base-quota delivery, had been eliminated.

Grain sold above the base quota is either a mandatory sale at the 50-percent bonus price or voluntary at negotiated prices. Negotiated prices are not necessarily higher than the above-quota bonus price, and must be below free market prices in the area. In 1977, just over 0.2 mmt, or less than 1 percent of state grain purchases, were at negotiated prices.[8] In 1978, a good crop year, this increased to 3.25 mmt, and, by 1980, to 8.6 mmt.[9] In 1981, only 40 percent of state procurement of grain was at the base price, while 60 percent was at above-quota and negotiated prices. The low proportion of procurement at the base price was partially due to a failure of some teams to fulfill basic delivery quotas.[10] From 1979 to 1981, above-quota price and delivery increases constituted 27 percent of the growth in collective revenue, which raised peasant income by 8.55 billion yuan.[11]

The combination of higher purchase prices, over-quota sales possibilities and increased sales of grain by the government to rural areas has led over the past few years to greater agricultural commercialization. Price increases accounted for much of the income gain during 1979-1981, but further substantial price increases cannot be expected. Quota policy will be the key to further collective income growth, since over-quota deliveries were second only to total deliveries as a source of increased income by 1981.[12] The 1979-1980 income increase was in part generated by a fall in in-kind consumption in favor of commercial sales, which allowed income to increase at a rate above that of real output. However, Premier Zhao Ziyang has suggested raising procurement quotas and limiting subsidy payments on rural grain sales to help control budget deficits.[13] If implemented, these actions can be expected to slow the trend toward higher levels of commercialization.

COMMUNE AND BRIGADE ENTERPRISE

In 1978, commune and brigade enterprises generated 29.9 percent of total commune income, and 14.5 percent of peasant income from the collective sector.[14] Further development of these enterprises would clearly strengthen the collective sector. To encourage such development, a July 1979 regulation provided for two- to three-year tax holidays for new enterprises and a low income-tax rate for existing enterprises.[15] This regulation quickly led to vigorous development of some commune and brigade enterprises in lines competitive with state enterprises still subject to high taxes. Rather than accept revenue losses in state processing enterprises due to this new competition, the taxes on high-tax products were restored for all commune and brigade enterprises in January 1981, and they were enjoined not to compete with state-owned enterprise for raw materials.[16] A May 1981 regulation called for an end to the growth of commune and brigade textile, cigarette, and salt works.[17] Commune and brigade enterprise grew rapidly through 1981, but the effects of the 1981 tax changes were apparent as value added and net profits plus wages rose by only 5.7 and 3 percent, respectively, on total income growth of 12.5 percent.[18] As tax holidays for new enterprises expire over the next three years, profits will be further squeezed.

Commune and brigade enterprises contributed perhaps 11 percent of the total increase in peasant income between 1978 and 1981, rising to 16.2 percent of collective distributed income, but falling in importance to total income. The relative contribution can be expected to fall further in the next few years unless the present high rate of reinvestment of enterprise profits is lowered to permit greater distribution. It is likely, however, that the profits will continue to be used to finance collective production and welfare activities.

TAXES AND INPUT COSTS

The cost of production and collective withholding has an important impact on net peasant income. In 1978, decisions were reached to lower the agricultural tax and reduce input prices. The agricultural tax reduction is not designed to raise income generally, since it amounted to only 3.4 percent of gross commune income in 1978.[19] Rather it is meant to solve a longstanding problem with teams experiencing below-average harvests. Such teams fulfill their delivery quotas, but then have to buy back grain to meet members' subsistence food requirements. In 1977, 940,000 (out of 5 million) teams both sold and bought grain. The amount of grain repurchased was 2.3 mmt.[20] It is unclear just how much of this grain is actually delivered to the state before repurchase, but for that portion that is delivered the team must pay for hauling both ways and, in some cases, storage and management fees. During the 1950s, these fees may have ranged from 8 to 17 percent of the price of the grain, though a recent publication claims that management fees are now absorbed by the state.[21]

Relief from the financial and administrative burden of repurchase was given through an exemption from mandatory grain delivery for rice-growing areas with a yearly distribution to peasants of less than 200 kg/capita of food-grain, and other grain areas with distribution below 150 kg/capita.[22] A grain-tax reduction of 2.35 mmt, apportioned to the provinces, financed this exemption.[23] Provinces determine eligibility, standards, and exemption periods either yearly or for multi-year periods.[24]

The tax cut, and a better harvest, resulted in only 480,000 teams repurchasing a total of 1.25 mmt of grain in 1979.[25] Though the yearly value of the grain-tax reduction is less than 1 percent of 1978 collective income at current base prices, it could mean an increase of more than 5 percent of their collective income if targeted to the poorest 20 percent of China's teams.

In 1978, plans were made to reduce the prices of major manufactured inputs by 10-15 percent "due to reduced production costs."[26] Manufacturing costs were rising, however, and profit margins had already been reduced by price cuts over the previous decade. In the farm-machinery sector, the government faced strong opposition from bureau chiefs over the price-cut plan, and, by late 1979, had dropped specific price targets.[27] The price index of agricultural inputs rose 1.9 percent from 1978 to 1981, costing peasants about 0.65 billion yuan of net income at 1981 input levels.[28]

For peasant collective income, the price and quota changes in the marketing of agricultural output far outweigh the impact of tax or input price changes. The marketing reforms immediately boosted peasant income. They also created strains in the government budget, as purchase-price increases were not fully passed on to consumers. Official discussion of price and quota levels makes it clear that the government will not consider substantial price increases or quota reductions in the near future. To maintain the momentum of income increases, then, other policy innovations were needed. One of these was to optimize the distribution of crops.

The collective sector was yielded modestly increased power over cropping decisions in 1979. Previously, state control of cropped acreage had ensured allocation of most land to grain, in line with the slogan "Grasp grain production." This policy was implemented even in cases where land, due to physical characteristics and relative prices, was more suitable for the production of alternative crops. The allocation pattern was reinforced by a decline in traded grain from the 1950s to the 1970s.[29] During the Cultural Revolution period, the reduction in state-traded grain, coupled with a prohibition on free-market grain trade, effectively prevented production units from circumventing the acreage allocations. In 1978, it was decided that, on the premise that production teams were "accepting the plans and directions of the state," they had "the right to grow plants in a manner and timing suitable to local conditions."[30] This was not a manifesto for free cropping patterns. Though the modernization of agriculture was to be encouraged through specialization in production, this specialization would be carried out under the guidance of the state plan, not at the initiative of local production units responding to perceived market signals. Communes were still responsible for meeting state grain quotas and for supplying other crops determined by state contract. By the end of 1980, the policy was relaxed to allow acreage to be used for crops of the commune's choice if yield increases permitted quotas to be met on less sown acreage than in the past.[31] This has allowed a reduction in area sown to grain, as yields have increased since 1978.[32]

State-guided cropping changes seek to concentrate industrial crop production in the physically most suitable areas. Increased supplies of commodity grain and decreased grain delivery quotas have been used to induce the desired shift in cropping patterns, usually with the promise to exchange grain for the industrial crop at fixed proportions. In 1978, 1.8 mmt of grain were sold in this program; by 1981 the amount exceeded 5 mmt.[33] As a percentage of the grain crop purchased, in 1980 total rural grain sales were substantially above the levels of the Cultural Revolution decade, though still below the record levels of the 1950s.[34] Sufficient information is not available at this time to estimate the impact of cropping changes on total output and income. These changes may, in any case, be coming to an end; in 1982, prohibitions against further decreases in grain-cropped area were appearing.[35]

All the policy changes discussed thus far are consistent with a highly collectivized agricultural sector. However, in 1978, experiments began on means to heighten production efficiency by increasing the direct rewards for good work. The resulting changes in the system of labor management have created a substantial challenge to collective production.

LABOR MANAGEMENT AND PEASANT INCOME

Communist Party leaders recognized that the egalitarian distribution of collective income practiced during the Cultural Revolution period was a strong work disincentive for most peasants. In 1978, they declared that "'To each according to his work' and 'More pay for more work' are the socialist principles of distribution; it is absolutely not permitted to reject them as capitalist principles."[36] To implement these principles they suggested that "the commune organizations should fix labor quotas for field work, if this is suitable, and assign work on this basis. The field work done should be verified and evaluated. In fixing labor quotas and evaluating labor achievements, the quality of the work should be considered first. It is important to evaluate the work done by each commune member and calculate his work points based on the quality as well as the amount of his labor."[37] The problems of evaluation in such a system were what made the egalitarian approach appealing to cadres in the first place.[38] Though the Party continued to advocate use of this system through 1979, as early as 1978 Wan Li, the Governor of Anhui province, had permitted alternative systems which divided production teams into smaller work groups, or even households, which then took sole responsibility for a piece of land and were paid on the basis of its net output. By the end of 1978, over 15 percent of the teams in that province used such systems.[39] Neighboring Jiangsu province also began tentative use of the household responsibility system in 1979.[40] There was debate over whether the household and individual responsibility systems were socialist institutions, and they were not publicly approved until 1980.[41]

There are several variants on the group, household, and individual responsibility systems, which are detailed in Kathleen Hartford's chapter in this volume. In the one that appears to be most used, the group, household, or individual directly fills its prorated share of state delivery quotas, makes a specified payment to the team, and keeps the rest of the output. The peasants are, in essence, making rental payments on the land. In all the systems, inputs rationed to teams through the state supply system (for example, high-quality fertilizers) continue to be distributed through the teams, which also continue to own the land and to assign land use and quotas. Small farm tools are owned by the group or household; large tools may be collectively or privately owned. In 1980, 40 percent of the teams used group responsibility systems and about 20 percent household responsibility systems.[42] By the end of 1981, 81.3 percent of the teams used a responsibility system of some sort. Of these, three-fourths based production on household or individual laborers. Fully 38 percent of the teams had allocated all land, livestock, and agricultural implements to households, which were then responsible not only for the state quotas and team assessments, but also for generating their own investment funds.[43] By the end of 1982, 78 percent of the teams used some version of the household responsibility system.

Lack of data prevents evaluation of the output effects of the responsibility system, but it is clear that new management systems have succeeded in reducing the cost of production as a percentage of output. As Table 13 shows, less income has also been deducted for collective use since 1979, particularly for use in collective agricultural investment.

Cost of production as a percentage of output will probably continue to fall as the responsibility systems are more fully implemented, though absolute gains will not be great, as input prices continue to creep up. The efficiency effects of the responsibility system should be fully realized shortly after adoption, so upward pressure on production costs can be expected to reappear within a very few years.

The nature of the responsibility systems does suggest certain interesting issues. First, these systems privatize production and investment and blur the distinction between private plots and collective land. Land-use patterns will surely change in villages that adopt household or individual responsibility systems. Seccond, we can expect a change in the composition of agricultural

TABLE 13 Year-to-Year Changes in Collective Income and Costs, 1977–1981
(billions of yuan)

	Total Income	Cost of Collective Production	Collective Withholding	
			Total	For Investment
1977–78	13.15	3.90	1.21	1.16
1978–79	12.67	3.49	1.54	1.23
1979–80	1.80	1.26	-1.33	-3.14
1980–81	10.87	-0.62	-1.60	-0.74

Source: State Statistical Bureau, ed., *1981 Chinese Statistical Yearbook,* p. 195.

investment as the locus of investment shifts from the team to the household. This is occurring at the same time that national investment in agriculture is being reduced, thereby further altering the composition of investment. The change in investment locus, plus greater efficiency of input use, may explain the fact that real input purchases dropped in 1981 while output continued to climb. Third, the responsibility system substantially reduces direct collective, hence state, control of resources. The percentage of output going to collective uses has been falling and collective funding of social services is declining. This implies that access to education and medical care, to cite two examples, will become more dependent on family income. Fourth, the nature of resource-management problems has changed. Leaders must now deal with problems of land and commodity input allocation among households rather than with labor and input use over the land. Local-level irrigation-system management has also become a more important skill, because distribution disputes proliferate with the increase in independent users. Fifth, rural financial institutions face new challenges. Where the new systems are in use, loan distribution is to households, greatly increasing the administrative burden on banks. There is also much more cash in the economy, as teams had cleared most accounts by bank draft while peasants do so in cash.

Casual observation suggests that some analytical and management models used in capitalist countries are applicable to the issues highlighted above. It is important, however, to note that the marketing systems and ownership of land remain in state or collective control. Furthermore, the hiring of labor is closely controlled, as is rental of assigned land. The threat of restriction of private activity and re-collectivization of production serve to limit capitalistic tendencies in Chinese agriculture.

HOUSEHOLD SIDELINE INCOME

In Table 11, the striking fact is that household sideline income dominates income changes. Even in 1979, with the large price boost, increases in the collective sector just equal those in sideline income, and are below the total of the sideline and other income categories. The sideline income figures reflect increasing private plot size, and increased production and marketing freedom.

The legitimacy of private production and marketing was reaffirmed in late 1978. "Small plots of land for private use by commune members, their domestic side occupations, and village fairs are legitimate adjuncts of the socialist economy. It is not permitted to criticize them and ban them as capitalist" and, "While we consolidate and develop the collective economy, we must simultaneously encourage and assist the peasants to manage domestic side-occupations, thus increasing their personal income and invigorating the rural economy."[44] This official approval of private economic activity was much stronger than that given in early 1978, when such activity was still severely restricted. Flourishing free markets are the key to the peasant's ability to specialize in production either of agricultural products on the private plot or handicrafts in domestic side-occupations.[45]

In 1978, it was specified that "private land will usually account for 5 to 7 percent of the cultivated land owned by the production team, and this percentage may not be expanded or transferred."[46] This regulation on size, unchanged from the Cultural Revolution period, was soon challenged at the provincial level. Research in Sichuan province showed that crops produced on private plots had yields more than twice those on collective fields, and, in 1979, the provincial government decided to increase the private-plot limit from 7 percent up to 15 percent of cultivated land.[47] By the end of 1980, over 9.5 percent of Sichuan land was in private plots, though it was early 1981 before the national government formally changed the 7-percent restriction to 15 percent.[48] Nationally, cultivated land in private plots went from 5.7 percent in 1978 to 6.2 percent in 1979, 7.1 percent in 1980, and 7.5 percent in 1981.[49]

Not only was private plot size increased, but restrictions on how the land could be used were eased. Cultural Revolution period restrictions varied by province and over time, but included limiting production to self-supply, restricting animal types and numbers, and placing limits on income. These were generally lifted by the latter half of 1980.[50]

PRIVATE MARKETING

The increase in free markets and their total trade was extraordinarily rapid. The number of markets is said to have increased substantially in 1979, and rose another 36 percent to 40,809 by the end of 1980.[51] Some 5 mmt of grain were traded on these markets in 1980, increasing marketing of domestic grain by 8 percent above the level of government purchases.[52] Total free-market sales rose from 18.3 billion yuan in 1979 to 28.7 billion yuan in 1981. Sales to urban residents, an important indicator of marketing freedom, nearly tripled between 1978 and 1981, from 3.11 to 8.94 billion yuan.[53] The expansion of private production and marketing made it the fastest growing component of peasant income by 1980.

Free-market prices have stayed firm in the face of increased supply, rising 6.7 percent in 1981.[54] However, to the extent that peasants buy as well as sell on the free market, price changes will not necessarily result in real income changes for the sector as a whole.

When considering the future path of income from this source, the implementation of the production responsibility systems becomes important. From the perspective of production management, in the purest form of the household responsibility system there is no difference between the private plot and collective land managed by the household. The productivity advantage previously found on private plots should therefore end. Of course, under management systems akin to this one, the difference between collective and sideline income is itself unclear.

OTHER INCOME

Other Income covers numerous items, but the principal one is remittances from off-commune employment. Capital-construction projects are a large employer of unskilled commune labor, so total capital-construction investment is an important determinant of this income. Other policy decisions, such as the attempt to free jobs for unemployed urban youth by forcing people with commune legal residence out of urban employment, have episodic influence. The 1981 increase of 8 million commune members taking part in collective distribution, after two years of no growth, probably reflects these factors.[55] Increasing income from petty trading and other private occupations will gain importance, but the central government failure to restrict investment to desired levels is the best explanation for increases in this income.

Table 11 shows net per capita real income increasing at a rate of 11.4 percent per year from 1978 to 1981, with year-to-year increases of 14, 9.4, and 10.7 percent. Before commenting on the sustainability of this rate of increase, some income-distribution implications and consequences will be examined.

INCOME DISTRIBUTION

Available information allows only superficial analysis of the intra-rural income distribution effects of the policy changes. A Chinese survey shows that, in 1980, the 27 percent of China's brigades with less than 50 yuan per capita distributed collective income derived only 18.6 percent of their total income from commodity sales; the rest was consumption-in-kind. The 9 percent of the brigades with greater than 150 yuan per capita distributed collective income derived 50.4 percent of their income from commodity sales.[56] An estimated 70 percent of the increase in gross collective income from 1978 to 1981 is attributable to price increases which, of course, cannot be captured without commodity sales.[57] The wealthier brigades, with in-kind distribution substantially above the margin of subsistence, have greater possibilities to substitute sales for in-kind consumption, as was clearly happening in 1980.[58] Income increases generated through the price mechanism will thus go disproportionately to wealthier brigades. The targeted tax cut mentioned earlier is, of course, a countervailing force, but is equal to only 7 percent of the first year price-induced income increase, and only 4.4 percent of the third year.[59] Within teams, implementation of the household responsibility system can also be expected to increase the skewness of income distribution. The egalitarian income-distribution method used previously tended to spread the return to exceptional farm management or other skills throughout the team. These will now be captured by the household alone.

The differences in urban and rural consumption changes over this period are interesting. The picture that emerges, shown in Table 14, is one of increased commodity consumption going substantially to the rural areas. At least through 1981, on a per capita basis additional output of consumer goods was largely captured by rural residents. This result can be shown through analysis of figures the Chinese have published on the retail sales of consumer goods.[60]

TABLE 14 Per Capita Retail Sales of Consumer Goods, 1978–1981

adjusted for inflation (yuan)

Year	Urban	Urban Including Free Market	Rural	Rural Including Free Market
1978	597.9	623.8	61.6	*
1979	523.8	558.4	73.4	94.5
1980	573.8	618.8	86.8	119.0
1981	575.2	630.1	97.0	136.9

Sources: Total retail sales of consumer goods by sector and sectoral population figures are in Guojia Tongjiju, ed., "Zhongguo jingji tongji ziliao xuanbian," pt. VIII, pp. 3, 23, 24; and "1949–1979 nian jingji tongji ziliao xuankan," pt. VI, pp. 3, 20. Free market sales are found in Travers, "Post-1978 Rural Economic Policy," Table A2.

This is a rather startling result. Despite substantial wage increases, urban consumption of purchased commodities increased by only 1 percent on a deflated per capita basis between 1978 and 1981. The effectiveness of the rural policy in raising rural purchasing power is evident, with consumption of commodities purchased from the state increasing by 57 percent over the four-year period on a deflated per capita basis. Rural purchases from state and free markets combined show an even faster increase.[61]

The inability of urban consumers markedly to increase their level of consumption is not due to a lack of buying power, as per capita statistics on personal savings show (see Table 15).

TABLE 15 Per Capita Personal Savings, 1978-1981
year end (yuan)

Year	Urban	Commune Member
1978	129.1	6.6
1979	157.5	9.3
1980	210.6	13.8
1981	255.3	19.8

Sources: 1978, 1980, 1981: Guojia Tongjiju, ed., "Zhongguo jingji tongji ziliao xuanbian," pt. VIII, p. 28. 1979: Guojia Tongjiju, ed., "1949-1979 nian jingji tongji ziliao xuankan," pt. VI, p. 24.

Over the same period, cash holdings increased from about 15.4 billion yuan to 30.0 billion yuan, or from 16 to 30 yuan per capita.[62] It is difficult to believe that the increase in the urban savings rate implied by the changes in personal savings and cash holdings is entirely voluntary. In the shortage economy that characterizes China, there are simply not enough attractive purchasing opportunities to induce the use of those savings. The fact that the increase in per capita consumption in the past four years has occurred largely in the countryside is strong evidence that the distribution system succeeds in moving commodities not to areas of effective monetary demand but to areas of effective political demand.[63] Had the cash income increases realized by peasants not been transformable into consumption increases, the income policy would have been ineffective. The policy has, however, succeeded in redistributing income to rural residents while protecting the real income of urban residents.

FUTURE GROWTH

The net increase in real agricultural output has been quite good—3.5 percent per annum from 1979-1981.[64] Part of this growth has been due to increased specialization in production, a process that official pronouncements suggest will be slowed.[65] Price increases and over-quota sales have also stimulated output. Purchase prices will climb very slowly in the near future, and quotas may well be raised. Given average weather conditions, however, it should be possible in the short run to maintain the recent growth rate. Continued reorganization of production management will lead to output gains, and filling the latent demand for modern inputs such as high-quality fertilizers will aid growth.

In the long run, inadequate investment in agriculture poses a major threat to growth. National government investment in agriculture has been cut substantially. Collective investment funds have dropped precipitously. In the production-responsibility systems, investment is left to the group, household, or individual. The government authorities seem to feel that peasants, with their recent income surge, have sufficient investment funds. Some types of investment are best undertaken at the household level, but in a tenancy system, which China is quickly adopting, there are strong disincentives for some types of investment, particularly those that cannot be recaptured in a single year or are immobile. The ill effects of tenancy on investment can be mitigated through long-term contracts, which are already in use in some places, but even those have limited value if peasants doubt the stability of the system. There are well-publicized examples of contracts being arbitrarily changed in mid-season.[66] Some problems are solved by maintaining the collective structure while allowing individual farming. For example, irrigation-system management and upkeep can be financed through team levies on households. Financing of new projects may be very difficult, however, especially when expected benefits are variable across team land.

On the national level, the composition of agricultural investment may be more important than its magnitude. Chinese studies have shown that large-scale water-management projects, which heretofore have absorbed the bulk of national agricultural investment, have low utilization rates because secondary and tertiary delivery systems are not functioning well or do not exist. For such projects, potential for gain rests with prefecture, county, or commune-level investment. The important area for national investment is in agricultural science, experiment stations, and agricultural extension. The positive externalities that make such investment so socially profitable are well known, and are the reason that government investment is essential. These investments have a long gestation period, but, with the labor intensity of China's agriculture, biologically based output advances seem necessary to sustain output growth at a level near that recently enjoyed.

CONCLUSION

The 1978 agricultural policy decisions were not uniformly implemented. The package of policies involving the government budget has been quietly repudiated, while other policies, notably those on sideline production and marketing and the management of collective labor, have been reformulated in ways that have fundamentally altered the rural economy. It is obvious that the architects of the 1978 policy were not motivated by a clear vision of the optimal structure of agricultural production. The production responsibility system, for example, was being experimented with in at least two provinces in 1978 with blessing from the central leadership, but it was not until 1980 that the experiments were judged successful enough to warrant widespread adoption. Even then, efforts were made to keep the production process collectivized in relatively high income units, a policy not relaxed until late 1982. New problems created by the responsibility system include land use, investment, credit, and resource management.

The increase in income from sideline activities, combined with other non-collective income sources, has led to a loss of importance for collective income. The collective-sector contribution was 58 percent of net total income in 1978, but dropped to 49 percent by 1981. If income from household and individual responsibility systems could be disaggregated from collective income, it is likely that under one third of total income would be found to originate in collective labor activities. The percentage will continue to fall with the widening implementation of household responsibility systems. While the "Other Income" source may fall if capital construction is cut back, sideline income should continue to grow as long as free markets are not further restricted. The current Chinese leadership has decided that increased agricultural output and peasant income are more important objectives than socialized production. The results of their policy changes suggest that, at China's stage of development, collectivized production had created substantial inefficiencies inhibiting the rapid growth of agriculture.

CHAPTER FIVE

Politics, Welfare, and Change: The Single-Child Family in China

JOYCE K. KALLGREN

Despite the current emphasis on stability, the so-called Single-Child Family (SCF) birth-planning policy must be considered among the more revolutionary policies introduced into post-Mao China. It is revolutionary because the leadership is attempting to bring about compliance by rural families with a new cultural norm of a single child for each couple. This flies in the face of a long tradition of values calling for many children (especially sons), a welfare program in which children form the major bulwark against the hazards of old age, and an economic boom that rewards families who have additional laborers.

It is not a policy adopted with great enthusiasm. As the *People's Daily* acknowledged, "The target of optimum population can only be achieved with the persistent effort of several generations. The policy of one child per couple is certainly not a perfect one. However, it is the only practical and provisional way within a certain period to shift our country's population from the situation of unhealthy development to the tack leading to an optimum population."[1]

In contrast to most other social-change legislation that has been adopted in China since 1949—such as the first of the two marriage laws (1950) and labor insurance regulation[2]—the Single-Child Family has not been set forth in national legislation. It was mentioned in the second marriage law; and family planning is explicitly made the responsibility of citizens in the 1982 PRC constitution. Though there has been discussion of a birth-planning law, the Sixth National People's Congress did not pass any such legislation in 1983.

The delay in formulating legislation can be attributed to controversies over the nature and scope of incentives and penalties, worries over the military consequences for recruitment in families where there is a single child, and general dissatisfaction with the policy itself. By mid-1984, there were those who still urged a national law, arguing that such a provision would insure equity among the various provincial and local regulations. It is unclear at the time of writing whether a national SCF policy will be adopted or, if so, what shape it would finally take. Existing policy may still be modified to alter its effect while retaining the original intent.

The possible components of the program, its major provisions, problems implicit in its administration, the rate of success or failure, and the likely direction of future developments have all captured the interest of Chinese leaders and outside observers. Some preliminary analyses have been published,[3] in which certain themes seem to recur. First is the effort to show how the SCF policy works in a specific locale through a combination of ideology, practical measures, and penalties. As with much of the successful social-change legislation enacted in post-1949 China, efforts to alter conduct combine propaganda and practical policies. In the case of the Single-Child Family, one problem has been the inability to create a situation in which the general population supports the policy, with only a small group of people perceived as being in opposition. A second problem has been the emergence of basic conflicts between equally valued goals of the SCF policy and agricultural development. The conundrum this conflict presents for modernization efforts is discussed in more detail below.

To understand why so revolutionary a policy should be adopted in the face of these quite serious problems and significant departures from preceding patterns requires some discussion of historical precedents. This chapter begins by briefly delineating the major features of social-change programs in the post-1949 period and the family-planning efforts that preceded adoption of the SCF policy. It is also necessary, however, to understand the current context in which the SCF has developed; thus, the following section outlines the agricultural changes that cradle the stringent SCF policy. Discussion of the major components of the SCF policy up to late 1983 and appraisal of the strengths and weaknesses of current programmatic efforts form the major focus of this chapter. In contrast to many of the articles and media accounts, the emphasis here is upon the rural—not the urban—sector. Evidence suggests that there has been much more success in meeting goals and developing a modicum of support from urban residents, but, since the great bulk of the Chinese population is rural, the success or failure of any birth-planning policy must ultimately be measured in the countryside.

HISTORICAL PRECEDENTS AND PATTERNS: SOCIAL CHANGE AND WELFARE

The Chinese leadership has been remarkably successful in undertaking programs effecting considerable social reform. The initial Marriage Law of 1950 required substantial changes in social practices. The law regulating labor insurance was complicated and difficult to establish but enjoyed the obvious support of many individuals.[4] The effort to outlaw the use of opium and suppress prostitution (also a feature of the early 1950s) has generally been regarded as successful, not only by the Chinese but by many observers.[5] Programs of social assistance for those in need—such as the Five Guarantees for rural families—have been well received and carried forward. Nationwide efforts to improve health and offer universal education have generally been greeted with enthusiasm.

Successful programs have shared certain features, regardless of the problems they sought to address. First, each has customarily been initiated by nationwide propaganda that explained the need for the program, the problems it addressed, and the value to the nation as a whole. This propaganda also made a conscious effort to isolate and label, and thus to define sharply, those who might oppose the effort. Whether the purpose was to bring about wholesale performance of a particular course of action, or avoidance of proscribed conduct, the Party and bureaucracy attempted to separate off the small group of those in opposition from the uninformed, uneducated, and perhaps unwilling participants in the criticized activity.

A second common feature of the various programs has been the establishment of a set of incentives available to those supporting the course of action. This has been coupled with penalties for those who are not sufficiently persuaded by education and reason to adopt the required values or course of action. Examples suggest how this has worked in practice. When labor insurance was introduced, the leaders wished to encourage labor-union membership and so increased benefits for those belonging to trade unions.[6] In the efforts to end prostitution in the city of Shanghai, those who came forward to confess their guilt were treated leniently, and often were provided with alternative employment; those who were hesitant or unwilling to admit their guilt were punished more severely.[7]

A third feature of social-change efforts has been a sensitivity to local conditions. National legislation is locally interpreted with the expectation that cadres will take account of economic conditions, the presence of minorities, or other special conditions existing in the locality.

In all these efforts, both positive and negative, tension is reduced by a series of social-welfare provisions. Naturally, the state has coercive power at its command and citizens know that the power is used occasionally, but education and persuasion (occasionally applied forcefully) are seen as the more important factors in bringing about compliance. While the SCF policy shares these characteristics with earlier Chinese reform programs, in this case it has proved exceedingly difficult to: (1) establish a small group of transgressors; and (2) allocate sufficient resources for use in rewarding adherents.

PLANNING A BIRTH-CONTROL POLICY

It is not altogether clear what set of circumstances brought about the decision for the SCF policy. Despite obvious statements about rapid population increase, and the evils of the Gang of Four, there seems little question that progress in reducing the birth rate had been achieved by 1979. This is true even though the history of population growth and family planning in China has been a checkered one. The Chinese leadership has vacillated in a controversy over whether to emphasize "many mouths or many hands." Although initially a drain on resources, children and a youthful population could be expected, at some point, to shift over into productive labor. Should policy stress the costs to society during their period of dependency, or the productive contributions that would follow? Changes in emphasis have caused the Chinese experience with birth planning to shift back and forth between restrictions on population growth and a more or less laissez-faire approach, which has been interpreted in the countryside as disinterest and perhaps as tacit approval of individual family choice. It is fair to say, though, that it was not until "after the Cultural Revolution . . . especially after Mao's death and the demise of the 'gang of four' in 1976 that the size of the population and its growth assumed significant priority in China."[8]

Quite apart from the matter of political values and doctrines, a realistic Chinese birth-planning policy cannot be separated from the development of sufficient scientific knowledge and technological resources to make generally available the means for preventing pregnancy and childbirth. It was only in the 1970s that a sufficient number of devices, pills, and other means came to hand to implement a universal birth-control policy. Even in the 1980s, reports persist of areas where abortions remain the only means for keeping the area within the target figures.

Between 1971 and 1979, a birth campaign was in effect

(launched by State Council Directive No. 51 issued by Zhou Enlai).[9] It called serious attention to the matter of controlling births and was based upon three policies: later marriages, longer spacing between children (a period of four years was commonly recommended), and fewer children. At the same time that this policy was widely endorsed, the capacity of China to produce the necessary pills and devices expanded. Even more important was the dramatic increase in the number of trained medical workers capable of inserting IUDs (generally prescribed as the means for rural women, since they might have difficulty following directions for prescription drugs).

The policy was rationalized largely in terms of the health and well-being of parents and children. If men and women married later, their educational and technical careers would be more developed. If children were better spaced, the health of both mother and child would likely be better. The opportunities for development of both would be enhanced, and the loss of work time would be reduced. Although the Marriage Law of 1950 set 18 years for women and 20 for men as the legal age for marriage, cadres encouraged later marriage. Sometimes they would not permit registration until ages 23 and 25. The effective marriage age seems to have been two to four years later than the legal age.

It seems likely that the policies were more effective in the cities and suburban areas than in the more remote countryside. Careers, alternative employment possibilities, educational opportunities, and similar matters—together with the capacity to bring pressure to bear upon fertile women—were all more effective arguments to use in urban environments. The situation in the countryside still could provide support for the policy of delayed marriage and fewer, better-spaced children. But the economic contributions of children to the collective family budget made the matter of child-rearing a much more controversial matter. Once the dependency period of children had passed (and some would argue the addition of another child need not be terribly expensive), the whole family could profit from the increased income collected when all members of the family were fully engaged in productive labor.

Statistical data to measure the situation prior to the adoption of the SCF policy are not plentiful, but, according to a Chinese researcher, "since the early seventies the population control work in our country has achieved major accomplishments. Within the brief span of just over ten years, the amount of new increase in population has fallen from 25 million to 12 million per annum. The natural growth rate of the population has decreased from 2.94 per cent in 1970 to 1.2 per cent in 1980."[10]

If this statement is true and a degree of success had resulted from these early Chinese efforts, it makes the adoption of the more problematic Single-Child Family program all the more puzzling. After all, the successfully declining birth rate would seem to suggest that China could eventually reach zero population growth. Moreover, the existing policy must have had some rural support or the national figures could not have shown the decline they did. In the rural sector, argue Parish and Whyte in their insightful work,[11] the small communities of peasants would not willingly oppose, by means of pregnancy, the group targets assigned by the authorities; such action might have negative consequences for other couples. Thus, social pressure coupled with the positive arguments about health, and the feasibility of a program based on later, fewer, and better spaced children, resulted in progress overall.

Why, then, did the leadership insist upon the SCF policy? The answer lies partially in the acceptance of far-reaching modernization goals for the year 2000, and what the leadership believed were the harsh realities necessary for their achievement.

THE FOUR MODERNIZATIONS AND THE PRODUCTION RESPONSIBILITY SYSTEMS

The period after the death of Mao and the removal of the Gang of Four was filled with recrimination, a litany of evils attributed to the imprisoned Maoists, and efforts to overturn verdicts and, where feasible, to return to power former Communist Party members treated harshly during those ten years. There were also, however, more serious discussions of the state of Chinese modernization, particularly recognition of the lack of progress in rural production. The period emphasizing collectivity had witnessed little if any growth in peasant income (see the chapter in this volume by Lee Travers). Within the commune, family income was based largely upon the work-point system (with a settlement of accounts at harvest time); the incentive or interest of peasants in raising productivity levels was judged to be diluted by this arrangement. Since the work points were assigned more or less equally, there was little advantage and even some danger in exceeding performance goals. Outside the commune, the private sector was limited to the so-called small plots assigned to each commune member where the family grew vegetables or raised a pig or two to supplement their income. All this outside work had to be undertaken discreetly, since to be charged with harboring capitalist impulses would provoke criticism. The private plots were deemed necessary but not permanent.

All aspects of rural modernization and production came under review in the post-Mao period. After an initial period (when some rather unrealistic goals were published and then set aside for more realistic and long-term planning), the so-called Four Modernizations—that is, agriculture, industry, science and technology, and defense—were set as goals to be achieved by the year 2000. These terms were refined and cast into specific targets that were ambitious yet feasible. Moreover, the goals were also defined in terms of personal income, requiring an approximate yearly growth rate of about 8 percent during the 1980s and 1990s.

A key part in this effort was to be played by agriculture and the raising of rural production levels. To that end the government and Party experimented with ways of developing and enhancing the enthusiasm of Chinese farmers, most notably by the introduction of production responsibility systems (see the chapter by Kathleen Hartford). The purpose of these innovative agricultural experiments (as well as comparable ones for the cities and their factories) has been to support China's rush to modernization. And as the Work Report of Premier Zhao Ziyang to the Sixth National People's Congress amply demonstrated,[12] there has been growth in the agricultural sector. The obvious satisfaction brought by this progress, however, has been accompanied by the problem of the consumption of state (and individual) profits and funds by an increasing population. In essence the Single-Child-Family policy is a reflection of the fear that the achievements from new productive policies will be wholly consumed by the rising population. Arguing that, since the population is still young, with 65 percent under 30 and hence in the critical childbearing age group, the regime fears that goals for the year 2000 would be in jeopardy if the population were to continue to grow, soon exceeding the goal of 1.2 billion. If, on the other hand, that figure can be sustained, then the programs of the Four Modernizations have a reasonable chance of achievement.

It is not difficult to understand that a burgeoning population would place extra burdens on the government, nor that the leadership might well wish to perpetuate, even underscore, the policies of the early 1970s. What is not clear is how the maintenance figure of 1.2 billion was established. Even more important, and also unclear, is how targets for yearly birth rates were established, or how allied procedures were designed for predicting population increases.

A population study cited earlier includes a useful chart that incorporates what is known about the birth-planning organizations in the national government, coupled with indications of the equivalent organizations at provincial and lower levels.[13] These charts and the organization to which they refer, cited in various speeches of Chinese officials and in the myriad references to the 1.2-billion goal and its importance for the Four Modernizations, do not fully indicate the source of projections used for recommendations from the State Council. However, it seems that the statistical analysis underpinning these data is derived from projections based on a series of different fertility patterns, depending on the number of children produced per couple.[14]

Momentarily setting aside the matter of statistical reliability, and the basis upon which the computer projections are made, what is most significant about the approach is the fact that the arguments for social approval are quite different from those that obtained between 1971 and 1979. Despite an occasional cartoon, poem, and the like extolling the value of a single child,[15] the thrust of the argument for the SCF is based on social needs and national interest. The effort to sharpen the rate of decline is a national-level decision. A reduction in children

TABLE 16 Chinese Population Projections for 2000 and 2080

Projections	Number of Children per Couple	Projected Size of Total Population (in 100 millions)	
		Year 2000	Year 2080
1	3	14.14	42.64
2	2.3	12.82	21.19
3	2	12.17	14.72
4	1.5	11.25	7.77
5	1	10.50	3.70

Source: *Population and Birth Planning in the People's Republic of China*, Population Report Series J no. 25, January-February 1982, pp. J-581.

born will reduce the need for services and the conflicting demands on profits and outputs. These profits will then be available for modernization efforts and to improve the lives of individuals.

Thus, the government has adopted and developed a policy with the expectation that a certain fertility outcome will result. In this process we are seeing the reverse of what demographers are currently willing and able to do. As demographer Geoffrey McNicoll notes: "It is a fairly simple matter to show how a given pattern of fertility 'makes sense' in a particular social and economic setting. Knowing the answer to a fertility outcome, the problem is essentially to infer a set of weights that applied to different institutional forms in a society can recreate an incentive structure consistent with the outcome."[16] This incentive structure comprises pressures directly or tangentially bearing on fertility and can include economic incentives, legal or administrative sanctions, or social pressures on individuals to conform. What is not so clear—and what is being attempted by the Single-Child-Family policy—is how to affect the problem from the reverse end. That is, how does the state establish a set of structures that will result in a desired fertility rate?

Statistical reliability plays an important role in analyzing Chinese decisions, whether one is discussing the likelihood of reaching a specific target or, in the case of the SCF, avoiding an excess over a set figure. The initial decisions to execute the SCF policy were made before the 1982 census; they were especially subject to the weakness of data.[17] Chinese demographers have been very sensitive to these problems and wary about the manner in which unskilled commentators might seize upon portions of the report without understanding the full range of implications. (An example may be found in *Guangming ribao* of 12 January 1980,[18] where one Chinese cadre's proposal for "negative population growth" was rapidly denounced by scholars, who pointed out that the author of the proposal—although possessing strong political commitments—did not fully understand the consequences of the policy he was advocating.)

Despite the cautionary import of the above comments and the difficulties of data problems in China, demographers have begun to have some confidence in their knowledge of fertility levels, leading one Western scholar to comment:

The birth rate of 18.3 (in 1978) and the implied total fertility rate of 21.4 may thus be understatements. But the approximate consistency of the various figures (imperfect though we note them to be) including the age pattern of fertility support a belief that a remarkably low level of childbearing has been reached in China.[19]

One more comment about the agricultural responsibility system and the population projections must be made before turning to the SCF policy itself. Both the reproduced chart and a research report by Liu Zheng, Wu Canping, and Lin Fuda entitled "Five Recommendations for Controlling Population Growth in China" (written in 1979 and including a section entitled "Several Estimates on the Future Prospects of our Population Development") lead the reader to believe that, despite the stern warnings of the Chinese press and many Chinese leaders, the planning estimates of the Chinese demographers project a much lower level of compliance with family planning than might be expected from a reading of the media.[20] It is important to keep this fact in mind when considering the SCF program and, particularly, the difficult choices it appears to place before citizens and leaders. Later we shall suggest that the government's decision to intensify sanctions may be related to the planners' best judgments about likely success rates. For the present, though, we focus on the rural dimensions of the Single-Child-Family program itself.

THE SINGLE-CHILD FAMILY IN RURAL CHINA

Regardless of the problematical data, the Single-Child-Family policy has been publicly adopted. The main concern in this chapter is with policy and its ramifications at the local level in China's countryside. Although only minimal attention will be paid here to the role of the central government authorities and the provincial offices, it is necessary to look briefly at the way the problem has been posed at the national level and the manner in which the target figures for birth planning reach the local levels, since this forms the backdrop to local efforts and provides a means for gauging whether or not a program is successful. Furthermore, the national statement of the problem gives some indication of the extent to which the CCP and the government will bring pressure to bear on the local authorities for effective implementation of its policies.

In its simplest form, according to Chen Muhua, "The immediate primary task is to advocate that each couple have just one child; the resultant problems are secondary."[21] Although the quotation is drawn from a speech by the Vice-Premier, the language is that of Chen Yun, a senior statesman closely allied with Deng Xiaoping and primarily an economic planner. Both Chens recognize that a wide range of support measures must accompany such a policy. These include the need to establish incentives to reward those who comply with the goals; to promote health care and medical facilities for pregnant women; to insure that a child is successfully carried to term; to develop effective and safe birth-control techniques as well as procedures and practitioners to facilitate sterilization for those willing to undertake such a step; to establish a series of positive economic incentives to reward those who accept this policy; and to design effective long-range measures to meet the social-welfare needs of couples who accept the SCF policy and therefore might be expected to need financial and other forms of assistance in their later years.

The policy has been advanced since its adoption by propaganda campaigns; for example, January 1983 was designated a month to concentrate upon the SCF and birth-planning policy efforts.[22] There have been a succession of national and local planning conferences held by those charged with implementing various aspects of the program—such as the establishment of targets for yearly assignment of planned births[23] and the local development of specific programs for rewards and punishments of those supporting or violating the policy.[24] There has also been renewed attention to population quality. The 1980 Marriage Law, for example, contained restrictions on those who may marry. Furthermore, there has been continuing research on the means to reverse sterilization procedures when personal circumstances or national disaster make such a course of action desirable. Such research has been underway in China for some time but has received special attention in recent years,[25] prompted, for example, by the desire of many couples in the aftermath of the Tangshan earthquake to have another child when their children had been killed; or occasionally by the wishes of a couple to have another child after remarriage or the death of an infant.

But the bottom line of the Single-Child-Family policy is that "it is not yet possible to stop the population from expanding in the next twenty years. We can only control the birth rate and gradually lower the natural rate of increase in order to reduce the latter to zero by the year 2000."[26] That is, since many of China's young men and women will reach marriage age in the next decade, the government realizes it cannot characterize resisters to the SCF as a small group of dissidents, nor can it stop the population rise altogether. Concerned about these hurdles, and faced with the consequences of indecision and laxity with respect to population growth in years past, the government has reacted with a relatively harsh policy now.

To be meaningful, the policy had to establish the link between population control and the significance for individuals of the success of the Four Modernizations. This was expressed, initially, as a goal of achieving US $1,000 per capita (the amount has since been reduced, but the argument is still pertinent). The government could then show the impact on that figure of various population growth scenarios:

The Central Committee of our party has said that our gross national product should reach an average of US $1,000 per capita within this century. In order to reach this target, and if our population did not exceed 1.2 billion, then our GNP would have to increase at an annual rate of 8.6

percent, but, if the total population reached 1.3 billion, then the GNP would need to grow 9.6 percent. The latter rate would be very difficult to achieve.[27]

Yet, problems remain in the government's effort to establish the link between individual self-interest and national policy. These problems may be briefly stated as (1) contradictions between the interests of the individual and the state; and (2) contradictions among different policies of the state in their respective impact on individuals.

The contradictions between the individual and the state are perhaps the most obvious, for the Single-Child-Family policy must be viewed as a threat to the personal security, not only of the cohort of couples now entering into marriage, but also to the cohorts of their aging parents and grandparents. No matter how strong the reassurances made by the state, the need for assistance to the aged is a fundamental fact of life for China. Though progress has been made, the Chinese press acknowledges that "at the present stage our country has very little in the way of social insurance for the care of widows, widowers, orphans, childless persons, and cripples. Relatively speaking their life in their later years is not as happy or prosperous as that of people with children. This has great influence on most people, and they always want to raise several children when they're young to avoid being without any way to live when they get old."[28]

The problem of old-age assistance is exacerbated by the particular nature of family living patterns in China, which provide that the wife customarily moves away, quite often into the home of the husband's parents. This dominant pattern will, of necessity, undergo change in the event the Single-Child Family becomes a more common occurrence. While the 1980 Marriage Law provided legal encouragement, the social changes required are complicated. These might be expected to occur more commonly in the urban areas, where living patterns are already influenced by space shortages.

There is also the simple mathematical reality that a successful single-child policy could result in a couple's being responsible for the support of four grandparents and the single child. Since women make limited contributions to agricultural labor, this raises the possibility of a lone man assuming the major responsibilities for a large number of people. Government reassurances aside, this remains a perceived threat by all participants: the couple, the male worker, and the grandparents.[29]

Virtually all Chinese analyses of family planning recognize the need for increased welfare protection and facilities for those aged who have no children or only a single child.[30] All endorse state development of facilities. However, the economic realities of construction and the competing demands of other state programs clearly mean that an effective solution will be long delayed. Taking into account all these realities, public acceptance of the Single-Child-Family policy will require constant pressure by the state since the reward structure must, of necessity, be modest.

A second important contradiction between individual and national interests has developed because of important changes in rural production. As noted, a key feature of the contemporary modernization effort in China has been the emphasis on increased agricultural production and the increased freedom of the family unit. One consequence has been to heighten the contribution youngsters can make to their families' well-being by early withdrawal from school. A second advantage is gained from relatively early marriage and pregnancies.[31] Thus, as a result of policies that rely upon and encourage the confidence and enterprise of the peasantry, the family has re-emerged as a contributing economic unit; this development could not help but counterbalance the new emphasis on the single child. Chinese analysts certainly recognize the existence of this problem; as we noted, they urge careful propaganda work that will link family-planning efforts to the responsibility systems in such a way that the family accepts both family-planning goals and benefits from contracts for agricultural production.[32]

IMPLEMENTATION IN THE RURAL AREAS

Birth-control planning policy is tranformed from national goals to local decisions through the mechanisms of "birth targets."[33] These figures are determined at the national level and then adjusted downward for the provinces and further disaggregated into goals for rural and urban areas. Since the numbers of individuals marrying in any given year will vary, and since precise data for determination of local goals have not been previously available, it is understandable that national data are disaggregated into percentage terms measured against the rate of childbirths per 1,000. It is at this point that the rates for local figures are translated into figures that impinge on the individual couple. Some are given permission to have a child, some petitions are denied, and occasionally conflict arises when an unplanned pregnancy suddenly becomes apparent. Some units distribute the assigned figures to specific couples; some reserve the total for later assignment after the events.[34]

The SCF policy rests upon birth-planning procedures and mechanisms already in place in rural China. These procedures may be better developed and better organized in the cities, but they were available even in the countryside before this policy was adopted. Thus SCF goals can build on such procedures as the provision of birth-control devices without charge, the spread of knowledge about means for controlling births, and the maintenance of records on those women in their fertile years. Moreover, in the post-1978 period, family planning has been re-emphasized through high visibility accorded the national organization for the study of population and family-planning work.[35]

Various other developments have contributed to a declining birth rate. The specific requirements of the 1980 Marriage Law were designed to lower the likelihood of many births. It raised the legal age for marriage to 20 for women and 22 for men and stated clearly the family-planning obligations of all. Furthermore, the 1980 revisions called for a four-year interval between children.[36] All these provisions were adopted in light of the realities of Chinese population pressure.[37] They reinforce post-1949 values and those means known to contribute to population decline, such as delayed marriages and longer intervals between births (to reduce the total number of children a couple might have during their fertile years). But they do not necessarily lead to a single-child family; indeed, internal evidence suggests an underlying assumption that couples will have more than one child.

Most problematical about implementation of the SCF policy is the fact that the policy and its educational underpinnings must be restated anew year after year to mitigate (what the leadership assumes will be) continual pressure from each newly married couple. The SCF policy is carried forward by the birth-planning group which approves births for newlyweds (often the first to receive approval for a child), with the remaining priorities assigned, by the local authorities, to newly remarried couples with only one person having a child by the former marriage, or those where an infant unexpectedly died. Lower on the priority list would be those who already had a child and had waited the obligatory four years.

It is not illegal to have a second child, at least in the rural areas. Indeed, some local authorities may be considerably more permissive than expected, given national goals. In mid-summer 1983, the authorities in Guangdong were still announcing that two children were the norm in the rural areas. The conventional wisdom is "preferably one, two are all right, but never three." In those terms, the key emphasis has been placed on decreasing the number of births. As recently as 1980, Guangdong reported that 30 percent of the births in that year were third and higher order.[38] It is this figure that authorities are most anxious to reduce.

There are conditions under which a couple may seek to have a second child, although the list apparently differs from area to area. Ubiquitous to all is the birth of a non-hereditarily disabled child unable to be a productive laborer.[39] There is also provision for a second child for those in the more remote countryside who need the labor power, and those engaged in certain dangerous occupations. In addition, for minorities there has long been a recognition that their customs and cultural values might make them an exception to birth planning, though treatment of them has become increasingly strict as the SCF policy has been given more emphasis. Thus, there is a fairly broad range of circumstances that might permit a second birth allocation (though presumably the constraints of the figure allocated to the unit will limit possibilities).[40]

These exceptions assume that local authorities have the right to allocate birth permits, and that the peasants will recognize that right by their very act of seeking permission for a second child. Such a development institutionalizes the right of the state to be an arbiter in the matter of family planning; it thus replaces the family or lineage who might have been expected to determine these matters in traditional pre-1949 China.

The feasibility of sustaining limited allocations is enhanced by the development of a set of incentives to reward those who pledge to have only a single child—that is, those willing to announce publicly their commitment and to use either a birth-control device or undergo sterilization. The state for its part then awards a Single-Child-Family certificate to those making this commitment, presumably conferring a set of benefits on the single child and the parents.

These benefits are separate from those that accrue to an individual who submits to either an abortion or a sterilization procedure. The actual benefits for each are determined in the local area. The special rewards have included some award of work points to insure that the woman who undergoes abortion or sterilization does not lose work rewards when hospitalized or recovering; this was more common when work points constituted the principal form of economic reimbursement. Thus an abortion might result in a two-week grant of points, a sterilization in perhaps a one- to two-month award. Such an operation (especially sterilization) may also be rewarded with cash payments, a sewing machine, or some other remuneration.[41]

The rewards for child and parents in a Single-Child Family are more ongoing and presumably contribute to lessening the costs of raising the child. The local units customarily pledge to provide free health care, remission of school fees, often some monthly sum to help in child-care costs, and allocation of housing rights equivalent to a two-child family.[42] This last right is more precious in the city but still relevant for peasants living in communes in the peri-urban area where apartment dwelling is not uncommon.[43] The certificate may result in allocation of free plots equivalent to those held by two-child families, increased grain rations, and preference for commune enterprise jobs. In sum, the incentives may cover the range of rural economic benefits needed to replace the contributions made by the presence of a second child, but they are locally determined and locally financed. In addition to the rewards for the SCF couple, there is the national commitment to provide social insurance protection for those couples who forego the assistance that might be expected from the multi-child family.

The penalty structure is also locally determined, and varies. Penalties come into effect under two different sets of circumstances. For the couple that gives birth to a second child after accepting the SCF certificate, all awarded funds must be returned; this is followed by locally administered penalties. (One couple accepted the funds and placed them in the bank, in case they changed their minds.) With the birth of a third child, a couple is subject to financial penalties and the child, too, is deprived of advantages. The couple is usually subject to a multichild tax that is often collected on a one-time basis rather than through the years, probably because a monthly collection procedure would be too cumbersome.[44] Help in meeting various child-related costs—such as delivery, medical care, and so on—may not be forthcoming. There also may be penalties in the allocation of private plots. Access to child care and schools may prove difficult.

The discussion of penalties and rewards should not obscure the fact that coercion exists, some of it tolerable and some excessive. Couples with an unplanned pregnancy are subject to the often oppressive persuasion of mass meetings and personal visits to encourage "remedial measures." Public reports in the Chinese press, as well as the comments of scholars, indicate that local authorities are capable of bringing tremendous pressure to bear upon couples judged to be recalcitrant.[45] This is particularly the case with this policy, where there are national goals but little in the way of national regulations and rules. Evidence suggests that pregnancies are terminated as late as the third trimester with considerable danger and hardship for the couple, and anger and sadness for those forced to accept the procedures.

Attention and pressure are also being focused on those responsible for implementation, including the local birth-planning groups and the cadres charged with developing agricultural productivity and meeting local birth-planning targets. The first group has emerged from the growing emphasis upon improved medical treatment for those undertaking sterilization operations and abortions as well as those agreeing to insertion of IUDs.[46] Press reports often emphasize improved medical standards and have suggested that rewards be given to those who successfully and safely conduct such operations. Punishment is meted out to those who illegally remove IUD loops, presumably placing at risk the patients who undergo these procedures, and to those opposing policy.[47] Since many couples are not permitted to register the infant (and therefore to become eligible for grain rations) until a birth-control device has been inserted, the number of operations is high and likely to increase as the number of married couples increases.[48]

A second group involved in the process of controlling births includes the local committees that assign birth targets, grant exceptions, and integrate the birth-planning requirements with other agricultural targets. It is often the women who take on birth-planning tasks because of their gender, joined by some retired volunteers who visit fertile couples to ensure their knowledge of local regulations and the like. Often these individuals have little standing in the local-level bureaucracy.[49]

More important, as Chinese authorities have recognized, are the views and actions of those who constitute the local-level leadership. Their work is twofold: first, to serve as local models for the policy;[50] and second, to ensure appropriate support through various means, positive as well as coercive. Their success in meeting goals may result in bonuses and higher-level approval, or criticism and penalties in the case of failure.[51] The matter is quite complicated, since these local authorities have other demands for performance and also spend their lives in the local area, thus having to judge for themselves which set of constraints—job or family—is more important. The national leadership emphasizes the role-model position of local cadres, expecting that they will volunteer for sterilization campaigns and acceptance of the single-child limitation. Moreover, meeting the requirements of birth planning may result in a bonus,[52] the amount determined according to percentage allocations to various birth-planning tasks or through a "piece-rate" figure that assigns a small amount to each operation or survival procedure.[53] There are no data available on overall achievement of local-level targets. Reports of dissent resulting in demotion and other penalties are sufficient to suggest some measure of disagreement in local circles.

Implementation of the single-child policy has been dramatically affected by the full-scale agricultural changes of recent years. As families sign contracts to produce crops, goods, or services for a set price, the strength of the collective decreases. Rewards and penalties that once revolved about the work-point system are no longer effective methods for altering conduct. With the decline of the collective has come a decline in the financial resources available for social-welfare assistance not only to the so-called Five Guarantee families traditionally supported, but also as rewards for those who adopt the SCF policy. The problem has been only partially remedied by the insertion in the contract system of a tax to be paid for communal services. As the commune has declined in importance, its capacity to apply coercion has been limited by decentralization of the system. Moreover, the likelihood is that families with increased profits are less vulnerable to penalties imposed upon them if they ignore birth-planning restraints.

As consequences of the production responsibility systems have become clear, the procedures for limiting births have been altered. Authorities have combined an agricultural contract with a birth-planning provision: A family now must pledge to provide certain crops as well as to observe an agreed-upon birth-planning target.[54] For those unwilling permanently to accept the SCF policy, the alternative may be an agreement to accept birth-control devices for the life of the contract, running two or three or more years. Such a decision, though not a permanent one, permits both local authorities and the couple to sustain their relationship without friction. The couple may well hope and expect a change in government policy; the local cadre may find this degree of concurrence sufficient for the circumstances. The penalties for non-observance of the contract or unwillingness to take "remedial measures" can be inserted into such a contract system through restrictions on crops, prices, and related items, or withdrawal of lands, free plots, and similar items. From the viewpoint of the cadre administrators, the development of these interim procedures is attractive.

CONCLUSION: TOWARD NEW FORMULATIONS?

In October 1982, the initial figures on China's completed census were released, showing a population in excess of 1 billion. The population is still largely rural; those who are illiterate or semi-literate constitute more than a quarter of the population over the age of 12. Renewed statements about the necessity for ongoing family planning have appeared, coupled with thoughtful discussions of the procedures for implementing these efforts. As noted, 1983 began with a month devoted to propaganda efforts to raise the level of compliance with the SCF policy.

It is important, in considering the present family-planning program, to review the statistics presented in Table 16 on page 140. The projections indicate that the demographers' plans were based on a different level of compliance from that discussed by press, government, and Party spokesmen. It may be possible, therefore, to meet the 1.2-billion target with a much lower level of compliance than authorities are willing to concede, though one assumes that the leadership is aware of this fact in making their projections.

There cannot be any doubt about the national goals. Nevertheless, it is clear that the Single-Child-Family policy has run counter in both goals and execution to many features of previously successful formulas for social change. For instance, we noted above that, in earlier programs aimed at changing conduct, the majority of the population was presented as law-abiding and compliant citizens, with only a small "deviant" group of resisters; the government concentrated on reducing the size of this latter group. But the contradictory goals of current policies make it difficult to establish the conditions that would progressively diminish the size of the group latently or manifestly opposed to the new birth-planning efforts. This, too, has been recognized by the leadership; there have been numerous calls for a national family planning law embodying the legislative intent of the Single-Child Family, in the hope that this would reduce the difficulties of implementation by providing for well-publicized local regulations and regionally coordinated efforts.[55]

Another set of problems arises with respect to the terms of rewards and penalties. By and large the most effective implementation is to be found in the cities, where state control of housing space, and access to child care facilities, permits a more effective impact. But urban birth rates have been low for some time. The policies regarding fee remission and the like seem scarcely sufficient to determine the decisions made by most couples. With family units now of greater significance, fines and similar penalties are more easily established, but apparently offer fewer constraints if economic well-being is high. Furthermore, they are not easily collected.[56]

In the countryside, however, it is more difficult to establish and implement a set of positive inducements to reorient or reward approved social conduct. (One possible exception may be allocations of land for housing.) With respect to penalties, the withdrawal of medical benefits or increased costs of medical delivery and the like are only reluctantly applied by the Chinese, even by the strongest supporters of the limited-family policy, for they place at risk the health and well-being of the child and mother.[57]

In 1983, there was public discussion in China of the occurrence of rural female infanticide or disfigurement to permit the birth of a second child.[58] It is not possible to know whether this discussion reflects a decision to make public an ongoing problem (though reportedly infrequent), or the resurgence of a relatively rare occurrence under the pressure of severe and strict enforcement of the SCF system.[59] The discussion in the Chinese press and reports from limited interviews indicate that the typical conservatism of the rural community is reflected in this extreme response to an effective SCF policy. The press reports indicate that the government response has been imprisonment and a renewed educational effort. There are occasional comments suggesting that, in a few cases, the sex ratio in an area has been severely tilted as a result of local response to effective enforcement of a SCF policy.[60] This emerging information, when coupled with a more general resurgence of tradition in China (see the chapter by Elizabeth Perry) makes these accounts worrisome to Chinese authorities; it also indicates to the reader a strong public objection to the implications of the SCF program.

If we turn to the contradictions between state policies, we shall find some sharp clashes in goals. What is distinctive about the Single-Child-Family policy is that its implementation confronts the elite with competing claims and engenders instability. This is serious in the face of the considerable effort expended in the post-Mao era to convince the peasantry of the long-term stability of the leadership and its agricultural modernization policies. Yet, most of those policies have rewarded larger families. The upswing in production, especially given the limited investment of resources, suggests the success of those policies and the general satisfaction of the rural population. Perhaps most important, the new commitment to the production responsibility systems must be viewed as a conservative reassertion of traditional work patterns reminiscent of the old peasant tenant.[61] It is, therefore, not unexpected that the social practices that accompany this development revive, at least partially, older values and priorities. The family has been given and/or regained some of its traditional roles and, with that, some of its traditional mores. These economic and social patterns make the family-planning changes difficult to effect while the tools at hand are limited. It has been argued elsewhere that the Single-Child-Family policy has taken the national leadership into some of the most difficult areas of social change.[62] The continuing reluctance of local leaders to emphasize family planning reinforces this impression.

That change is possible cannot be doubted, as demonstrated by the experience in Sichuan and more especially the detailed report from Shifang.[63] But these cases also indicate that a significant portion of the achievements preceded the full-scale implementation of the production responsibility systems. Sanctions to force greater compliance from the rural sector are of course possible; the latent coercive power of the state is substantial. Its use incurs the risk, however, that agricultural growth would be compromised. Furthermore, the agents for implementation would of necessity be those leaders who already are charged with carrying forward other policies of importance. An increased involvement or diversion of their attention to family planning might well have unacceptable consequences.

At the risk of oversimplification, the major state argument

for the SCF policy rests upon an economic "balance sheet," namely that the costs to the state of a demographic explosion are too high, given other goals. In light of this assumption, it becomes important to look at the balance achieved. However, there is only fragmentary evidence about the amount being spent for the Single-Child Families either in a cash monthly payment or in services. We do not know how many people are receiving tangible benefits. Undoubtedly some areas are too poor to pay couples, and in some areas bureaucratic procedures hamstring efforts.[64] These matters aside, the system of incentives is designed to encourage people not to draw upon state resources through additional children. Yet, it is impossible to determine whether the current payment of benefits is, in fact, a smaller amount than might have been allocated as state support of a larger population. The argument will become more complex when and as the provision of social-insurance costs looms larger. We are certainly seeing a system in which local authorities are called upon to reallocate resources to those who have complied. Assistance to the single child is to be sustained to age 14. If one balances the costs of payments versus the alternative costs that might have been allocated to a family with more children, it is not at all clear what a comparison of the two "balance sheets" would show.

It may be necessary for the national leadership to attempt a new combination of policies. First, there would have to be a careful review and then institutionalization of practical suggestions for making more efficient use of the economic means at hand as rewards to those conforming to the preferred policy. At the same time, the public expressions of national goals would return to an emphasis on the more practical concerns about career and health benefits resulting from later marriage, and fewer and better-spaced children. Though emphasis on the single child would remain, it would be accompanied by a return to the issues previously emphasized, and more attuned to other individual goals and national policies. If the demographers cited here are correct, a slight birth increase may still not unduly compromise the Four Modernizations. The leaders of China have often reshaped their goals when the realities of policy administration have proved oppressive or when the costs have seemed excessive. Since the Single-Child Family presently appears to lack a number of features of earlier successful policy changes—isolation of a small group of resisters and allocation of sufficient rewards to encourage compliance—the leadership may well choose slowly to shift the emphasis of their efforts while retaining the slogans of the past.

CHAPTER SIX

The Implications of Rural Reforms for Grass-Roots Cadres

RICHARD J. LATHAM

When the Twelfth Party Congress was held in September 1982, various forms of the rural production responsibility system already had been practiced for nearly three years. Although the economic reforms appeared surprisingly durable in the eyes of some cadres, many rural members of the Chinese Communist Party continued to harbor reservations. One—albeit not the only—source of enduring criticism of the Third Plenum reforms between 1979 and 1983 consisted of the problems the reforms posed for grass-roots party cadres. The changes not only deprived grass-roots party leaders of prestige and political advantages, but also left cadres at an economic disadvantage and imperiled their normally secure leadership positions.

TWO CONFLICTING VIEWS

A careful review of Chinese commentaries, analysis, and scholarship reveals widely disparate views regarding the successes, failures, and implications of the rural production responsibility system. Most descriptions in the official press conveyed a favorable and largely uncritical perspective. *People's Daily*, for example, stated in July 1981 that "This kind of reform [that is, the rural production responsibility system] is imperative and irresistible."[1] Zhejiang's Tie Ying called the rural production responsibility systems an "igniter" of change.[2] On the eve of the Twelfth Party Congress, the China News Agency published extensive data about the successes of the rural production responsibility system in Guangdong. The breadth of the successes led the agency to proclaim that Guangdong was going through its "second golden age."[3] In yet another article in *People's Daily*, Dong Qiwu, a Vice-Chairman of the Chinese People's Political Consultative Conference, drew attention to the successes of the rural production responsibility system in Anhui and Shanxi.[4]

In contrast to such sanguine views, however, there also were references to difficulties, continuing disagreements, and even failures. In June 1981, *Hebei Daily* warned that, due to not "perfecting" the rural production responsibility system, there were "many cases of failure."[5] *People's Daily* voiced a similar concern one year later when it examined the case of a production team that had flourished under collective leadership but became a "grain-deficit team" after the rural production responsibility system was adopted. The problem, the paper argued, was not a result of the responsibility system, but of poor leadership.[6] One journalist observed in *Guangming Daily* that "there have been doubts and misgivings in varying degrees, and the problem in its totality certainly remains unsolved."[7] A similar attitude was expressed by Bai Dongcai, Executive Secretary of the Jiangxi Provincial CCP Committee and Governor of Jiangxi. Bai noted that "unanimous understanding has not yet been achieved." Some people and units, he said, were still "being reproached, censured, and suppressed" for adopting the rural production responsibility system. Some party cadres refused to include the adoption of responsibility systems on their work agendas. Such comrades, he concluded, "do not verbally protest," but they still believe that the rural production responsibility system constitutes "going it alone."[8]

The essence of the problem of persistent reservations about the rural reforms was summed up by Du Runsheng, Director of CCP Central Committee's Rural Policy Research Center. In a speech to the Twelfth Party Congress, Du observed:

> On some problems that everyone can sense, such as arbitrary and impractical directions, high procurement quotas, the price scissors in the exchange of industrial products for agricultural products, cancellation of personal freedom within the framework of the collective, and so on, *unanimity of views can be relatively easily reached.* For some other problems, however, the situation is different. For example, with regard to the management and administrative form of a commune or a production brigade—in other words, the so-called responsibility system—it is by no means easy to distinguish between right and wrong.[9]

Thus, at the same time that cadres could agree about the kinds of economic problems that plagued the rural collective economy, they could not agree about the fundamental appropriateness of the rural production responsibility systems. Reform advocates claimed their policies worked because they enlivened the rural economy and generally improved the economic wellbeing of the rural population. Critics of the reforms contended, however, that the reforms could be said to have "worked" only if important social implications and consequences were disregarded or de-emphasized. The differences between proponents and critics can be divided into three categories.

First, there were ideological differences of opinion. Critics claimed that the rural reforms represented a mistake in ideological orientation. They pointed to what they felt was an unmistakable departure from the socialist road. Well after the idea of rural production responsibility systems was introduced, some of the reformists themselves publicly criticized "dividing up the land," "going it alone," maligning the commune system, and establishing inefficient labor groups smaller than production teams.[10] In the main, defenders of the agricultural reforms claimed that the rural responsibility systems were simply new ways of "managing" rural labor forces." Ownership of the means of production, they contended, had not reverted to individuals.[11]

Second, critics and advocates of reform disagreed about the social consequences of the reforms. Critics argued that the reforms represented the abandonment of the socialist concern with equity issues. They claimed that the economic reforms would result in greater income disparities, the diminution of social welfare services, a waning interest in basic rural education, declining interest in military service, greater hardships for the poorest strata of Chinese society, a declining interest in public works projects, growing problems involving social disorders, and a deterioration of support for population control programs.[12] Conversely, defenders of the reforms claimed that "excessive equalitarianism" not only had failed to work, but had resulted in the perpetuation of poverty rather than prosperity. The "iron rice bowl" phenomenon (of guaranteed job security), they claimed, had been prohibitively costly and had stifled productivity and the efficient use of human and material resources. While acknowledging that some income disparities would result in the short term, proponents of the reform contended that the "rising river" of prosperity eventually would bring economic betterment to all working citizens.[13] Surveys were published to show that the poorest households in Chinese society had not automatically suffered from a loss of subsidies, income, or social-welfare services.[14]

The third category of criticism appears to have sprung from the managerial concerns of basic-level Party leaders. In general terms, grass-roots Party cadres objected to the reforms because they reportedly created chaos in the management of economic affairs. In the eyes of many rural cadres and peasants, the adoption of rural production responsibility systems implied there was no longer a need for production brigades and production-team leaders. Some local leaders also objected to new procedures which were only vaguely explained by higher authorities and were often poorly understood at all levels. Relaxing rural policies was seen as the primary cause of social and economic confusion in the countryside. And some Party leaders predicted that it was only a matter of time before retrenchment policies would be reintroduced.[15] The Party center argued, however, that the reforms created more, not fewer, responsibilities for rural cadres. If local leaders would use "flexible control," they asserted, chaos would not ensue.[16] The substantive challenge to local leaders, the CCP reformers advised, was to learn to use new leadership techniques and to realize the counterproductive consequences of coercion, commandism, patriarchism, and mass campaigns.

Underlying these ideological, policy, and management issues there were less defensible, but nonetheless widely held, reservations among grass-roots cadres. The reforms created anxiety among cadres concerning their future careers and their economic status. Generally speaking, China's rural population stood to benefit from the reforms, but there were few reasons for exuberance among grass-root cadres.

METHODOLOGY

A longstanding question among students of Chinese politics involves the relative importance of ideological imperatives and self-interest in determining political behavior. For much of the Maoist era, Western observers of Chinese politics found it difficult to detect anything but the pervasive intrusion of ideology in nearly all social matters. A particularly informative side effect of the Third Plenum reforms for China scholars is the new openness that is evident in Chinese publications. It is now possible to detect more clearly identifiable personal and organizational interests. Interest groups have always existed in varying degrees, but during the Maoist era it was hard to identify them with any clarity.

In the following pages, I shall seek to demonstrate how the Third Plenum reforms affected the personal interests of rural cadres. In particular, I shall examine the initial effects of the reforms on careers and economic status. The data were largely obtained from interviews in 1982 with Hong Kong residents who formerly were peasants and rural cadres in the People's Republic of China. Because all the respondents left China between 1978 and 1982, they usually were aware of how the rural economic reforms were implemented and how they affected various sectors of local society. Although 40 persons were interviewed, the observations of only 20 respondents were sufficiently detailed to be compiled in tabular data.

Certainly, the views of 18 peasants and 2 rural cadres do not constitute a complete picture of the diversity of cadre responses to the rural reforms. When the respondent data are used in conjunction with public sources, however, it is possible to describe general patterns and issues that existed not only in Guangdong but in Fujian, Anhui, and other provinces. Tables 17 and 18 illustrate the general demographic characteristics for the respondents' production units. Tables 19 and 20 summarize what the respondents perceived as some of the consequences of the economic reforms among basic-level cadres.

TABLE 17 Geographic and Demographic Data for Production Teams of Respondents, Guangdong, 1981

County	Commune	Number of Production Brigades	Teams	Team Population	Adult Workers	Arable Land (mu)
Chaoyang	Shitou	6	40	170	45	45
Chaoyang	Suchang	13	18	170+	50	50
Dongguan	Yizhai	13	13	154	–	–
Dongguan	Shagang	18	18	110	–	132
Huiyang	Zhaiyun	–	–	439	–	440
Puning	Changhe	5	10	100	–	10
Qingyuan	Foulin	13	12	210	50	245
Shenzhen	Shapu[a]	15	11	150	–	195
Shenzhen	Shapu[b]	15	11	118	–	180
Sihui	Malie	9	14	110	50	–
Wuhua	Jinshan	19	27	112	50	61
Yangjiang	Junyi	7	–	100	–	70
Yangjiang	Zhonglin	21	10	150	–	105
Zhongshan	Shanmei	15	10	245	130	303
Zijin	Yuling	15	14	140	60	60

Notes: [a]Dama production brigade. [b]Xiaoma production brigade.

By and large, the discussion of the impact of the reforms on the careers of basic-level cadres will be limited to leading cadres of brigades and production teams who held positions such as

TABLE 18 Geographic and Demographic Data for Production Teams of Respondents, 1981

County (Province)	Commune	Number of Production Brigades	Number of Production Teams	Team Population	Adult Workers	Arable Land (mu)
Tongan (Fujian)	Gangqiao[a]	—	—	—	—	—
Shou (Anhui)	Shiming	13	15	103	38	150
Jinjing (Fujian)	Dalieh	19	14	279	—	200
Guangrao (Shandong)	Laozhai	28	2	240	—	500
Daladqi (Inner Mongolia)	Hongtu	6	5	260	—	970

Note: [a]Gangqiao was an urban commune with comparatively few farm workers. The respondent was generally conversant about how the post-1980 reforms were carried out in Tongan County.

leader, deputy leader, bookkeeper, cashier, storekeeper, and work-point recorder. On occasion, I shall also refer to *non-leading* cadres such as barefoot doctors, agro-technicians, and teachers.[17]

"WORRYING ABOUT THEIR FUTURES"

Respondent interviews and Chinese press accounts both revealed that grass-roots cadres frequently viewed the agricultural reforms of the Third Plenum with less than enthusiasm. An important reason for this, observers claimed, was that the economic reforms created "worry and anxiety" among basic-level cadres regarding their careers as rural leaders. Some cadres, the New China News Agency reported in June 1982, "are in low spirits and lack sufficient morale." It was hard for them to concentrate on implementing the reforms because they "worry about their own futures."[18]

The CCP has published no figures that reveal the extent of "worry and anxiety" among rural cadres. We do know, however, that the size of the rural leadership elite (party and non-party

TABLE 19 The Impact of Agricultural Reforms on Rural Leadership Elites, Guangdong, 1980-1981

County	Commune	Cadre Commune	Reductions in Brigade	Team	Cadre Attitudes
Chaoyang	Shitou	NA	—	2/6	NA
Chaoyang	Suchang	NA	—	—	NA
Dongguan	Yizhai	NA	—	—	Did less CCP and cadre work
Dongguan	Shagang	NA	—	—	PB leaders did not want farm work
Huiyang	Shaiyun	NA	NA	NA	NA
Puning	Changhe	NA	NA	NA	NA
Quingyuan	Foulin	NA	4/6	3/5	Negative among commune leaders
Shenzhen	Shapu[a]	NA	—	3/7	Team leader supervises members less
Shenzhen	Shapu[b]	NA	—	-3	NA
Sihui	Malie	NA	NA	NA	NA
Wuhua	Jinshan	NA	NA	NA	NA
Yangjiang	Junyi	NA	—	—	NA
Yangjiang	Zhonglin	NA	4/6	-all	NA
Zhongshan	Shanmei	NA	NC	NC	NA
Zijin	Yuling	NA	—	—	Negative among retired PLA

Notes: [a]Dama production brigade. [b]Xiaoma production brigade.
NA = Data not available or unknown by the respondent.
NC = No change reported.
— = Reduction.
4/6 = Four of six cadre posts were eliminated.

cadres) is not small. Based on data supplied by the State Statistical Commission regarding the number of basic accounting units in the early 1980s, we may estimate that there were approximately 30 million leading cadres directly involved in the management of collective agriculture and enterprises at the grassroots level.[19] This may be a conservative estimate if the term "basic accounting unit" does not include production brigade leaders. In any event, the figure provides a rough idea of how many cadres were directly involved in agricultural management, at least at the production-team level in the early 1980s. More important, it was this sizable rural constituency that was

TABLE 20 The Impact of Agricultural Reforms on Rural Leadership Elites, 1980–1981

County (Province)	Commune	Cadre Commune	Reductions in Brigade	Team	Cadre Attitudes
Tongan (Fujian)	Gangqiao	NA	–V	–V	NA
Shou (Anhui)	Shiming	NA	5/6	3/8	More than 50% did not like RRS; did not like farm labor
Jinjing (Fujian)	Dalie	NA	–	–	NA
Guangrao (Shandong)	Laozhai	RRS was not adopted by the commune			
Daladqi	Hongtu	–	–	–	NA

Notes: NA = Data not available or unknown by the respondent.
— = Reduction.
5/6 = Five of six cadre positions were eliminated.
–V = Variable reductions depending on the kind of production management system in use.
RRS = Rural responsibility system.

charged with implementing the Third Plenum reforms and that stood to gain the least if not to suffer actual losses of political power and prestige. The incomplete data in Tables 19 and 20 suggest that as many as one third to one half of the cadre leadership positions in the units of the respondents were eliminated.

There were at least two general reasons for the existence of low morale among some rural cadres. These were the elimination of cadre positions and the problem of determining appropriate remuneration for the rural leaders.

Personnel Reductions

One objective of the Third Plenum agricultural reforms was to increase rural consumer income by decreasing production and administrative costs. This was to be achieved, in part, by eliminating the posts of some production-brigade and team cadres. There were frequent statements in the Chinese press which indicated that staff reductions in rural units were either necessary or already had been undertaken.[20] The observations of the respondents confirmed that such reductions had taken place in most of their units. Fourteen respondents reported the elimination of some leading cadre posts; only one respondent reported no change; and four did not provide any data. In the main, the units that adopted the system of households assuming responsibility for all production tasks (*baogan daohu*) were able to reduce the number of local leaders because there were fewer time-consuming responsibilities, such as supervising and organizing work tasks, calculating work points, and holding meetings.[21] Routinely, the positions of the deputy leader, bookkeeper, cashier, storekeeper, and work-point recorder were eliminated. Cadres and peasants alike often viewed the rural economic reforms as an indication that there no longer was a necessity for any leading cadres in production teams.

The responses of cadres to the prospects of fewer leadership positions were mixed. Many team cadres, for instance, believed that greater advantages were to be gained by resuming full-time farming activities. Taking on or continuing to carry out leadership responsibilities appeared less attractive than in the past. Attending to Party-building activities, for example, was also viewed as "disadvantageous" by some rural Party Workers.[22] In Fujian, a "new problem" developed at the grass-roots level: It became difficult to find people to fill leadership posts.[23] *Shanxi Daily* reported that, in the wake of the reforms, some leading cadres had "resigned their posts to return to their farmland."[24]

The existing data are still too fragmentary to demonstrate the extent to which the phenomenon of production-team cadres abandoning their leadership positions was a widespread problem. The respondents, for example, reported no cases of production-brigade or team cadres leaving their posts. They did note, however, that those leaders who continued to serve spent less time attending to collective affairs.

Brigade cadres appear to have responded differently to the staff reductions. Most brigade cadres were generally removed from direct involvement in routine agricultural work. Staff reductions for some cadres meant they would need to resume doing physical farm labor. Several respondents observed that brigade cadres were decidedly against adopting any household production responsibility system simply because it meant they would have to return to the fields.[25] One respondent claimed that the adoption of a rural production responsibility system was delayed as long as one year in his unit because brigade cadres did not like what the reforms portended for their futures.[26]

Some brigade and commune cadres who managed collective enterprises were also apprehensive. One target of the Third Plenum reforms was inefficient production facilities which were able to exist only because of the security provided by "iron rice bowl" policies. The implication of the reforms for some production-brigade enterprises was, for example, that they would have to compete for resources and markets with other commune, team, or even private undertakings. Unsuccessful enterprises would have to close. When one brigade leader learned that his brigade enterprise would be closed, he liquidated its assets and kept the money for himself. Appropriating production-brigade and team resources in the face of the radical changes apparently was not uncommon.[27] Following the adoption of rural production responsibility systems, brigade enterprises also were unable to "arbitrarily requisition" team assets and manpower as freely as in the past. Thus, brigade leaders had less maneuverability as local economic managers.

One journalist admonished rural cadres not to fret needlessly about the loss of their leadership positions. The reductions in personnel, he acknowledged, "will inevitably affect the jobs of some cadres." He continued, however, that "it is believed that the organizations will make proper arrangements for their placement." He offered well-intentioned but poorly informed words of comfort.[28] According to the interview data, as many as one in three cadres in brigades and teams may have had cause to be anxious about their jobs. Prior to the Third Plenum reforms, the leading cadres might have been able to lessen the impact of such personnel cuts by transfers, the creation of new party jobs, or by giving redundant cadres outright subsidies.[29] The insistence of the Third Plenum reformers that leading cadres must also have responsibility systems, however, meant that it would become increasingly difficult for leading groups to justify the maintenance of large staffs of non-production personnel. In Shifang commune, Wenxian county, Gansu, the positions of 178 commune and brigade personnel were eliminated. Cadres' subsidies, moreover, dropped by nearly 85 percent from 25,000 to less than 5,000 yuan per year.[30]

The Remuneration Problem

A second source of cadre apprehension involved new uncertainties regarding the basis of determining their income and benefits. Prior to the Third Plenum's agricultural reforms, loose financial control and management, particularly at the commune and brigade levels, gave leading cadres the use of commune resources without much public scrutiny or "democratic supervision." The point is not that cadres routinely engaged in pilfering, manipulating funds, or embezzling—although these certainly occurred—but that there was little insistence that administrative overhead costs should be commensurate with the productivity of each unit. In agriculture as well as in industry, the "iron rice bowl" syndrome made it possible to underwrite the support of large leading groups. Additionally, because the local party infrastructure had become increasingly synonymous with the management structure of communes and brigades, the income of collectives was also used to support non-production CCP personnel, party activities, and political movements.[31] Thus, in the absence of cost-accounting procedures, there had been no necessity to justify the overhead cost of the rural leadership elite.

The adoption of agricultural production responsibility systems altered the system of calculating income for cadres as well as for commune members. The changes appear to have varied depending on which system of production responsibility was adopted.[32] If a household quota system (*baochan daohu*) or "mixed production responsibility systems" were adopted, there continued to be a strong rationale for retaining a well-organized structure of collective leadership.[33] There was less necessity to carry out extensive reductions of cadre positions in these units compared to those that adopted the system of households taking full responsibility for all production tasks (*baogan daohu*). Production units that used a system of household responsibility for all production tasks required less direct supervision of agricultural work. There was also a greater incentive to decrease unnecessary management costs.

The income and related benefits of leading cadres were affected in several ways. First, the various production responsibility systems often resulted in what appeared to be a relatively fixed income for leading, non-production cadres at the same time that the income of peasant households could increase—often dramatically. Regular farm workers had control of production assets under the household production responsibility system, but leading cadres who were completely or partially removed from agricultural production had no such income-earning leverage. If their families were assigned responsibility fields or production contracts, their leadership responsibilities could seriously limit their contributions to completing household production tasks. In some cases, leading rural cadres who worked in commune or county organizations were not allowed to return to their homes to farm their families' responsibility fields.[34] In other instances, leading cadres were not allowed to sign contracts.[35] Cadres and their families were also admonished to avoid conflicts of interest and to place their economic needs behind those of regular commune members.[36]

One solution to this problem in communes, as in urban factories, was to encourage leading cadres to draft "responsibility-system outlines" (*gangwei zerenzhi*).[37] Once these job descriptions were prepared, the seemingly fixed incomes of cadres, which previously were based on work points, could be increased on the basis of superior work performance and the general economic success of their collective units.[38] Linking cadre income to the economic success of basic accounting units posed several problems.

First, it was difficult to establish such a linkage when the household production responsibility system was used. Increased peasant income did not necessarily result in immediate increases in the revenues of the brigades or teams. Compensation for leading cadres in some units took the form of slightly larger allocations of land, often without corresponding increases in land taxes or compulsory grain-quota sales. In one unit, the respondent reported that cadres received better quality land. Although allocating better plots of land to cadres represented one way of compensating them for their leadership work, the respondent viewed it as evidence of privilege-seeking.[39]

Second, it had proven particularly difficult to find meaningful criteria by which to measure the performance of non-production personnel.[40] In the countryside, as in urban areas, managers and bureaucrats were hard-pressed to demonstrate how their jobs contributed to productivity. Justifying bonuses and increased salaries for these people often was difficult at best.

Third, as commune members became increasingly conscious of production costs, assessments, and retained profits, they also became more sensitive to and critical of the visible cost of supporting the local leadership elite. In some areas, peasants refused to pay their obligations to the collectives. In Gansu, peasants called the high rate of collective grain retention to pay for the subsidies of local leaders and welfare services the "second lot of grain delivered to the state."[41]

Finally, the various rural production responsibility systems substantially altered the practice of awarding bonuses, subsidies, and other benefits which had long been used to supplement the incomes of leading cadres. For example, cadres and CCP members in one area were told to increase collecting subsidies for attending management and Party meetings.[42] Because the kinds of work that required supervision had decreased, there were fewer tasks through which some cadres could earn their work points. New accounting procedures meant that cadres had less access to the total output of the basic accounting units. There were, in effect, "slimmer pickings" for those leaders who were dishonest. The re-establishment of agricultural banks was followed by proscriptions that prevented cadres from borrowing funds from the accounts of their units. These economic consequences of the household responsibility system were not experienced equally by all cadres. For example, the respondent from Yuzhai commune noted that leaders of the inactive brigade militia still received subsidies from the brigade, even though they had been inactive for nearly two years.[43]

It appears that the phenomenon of leading cadres resigning their positions and potential leaders refusing to fill vacancies was a result of the perceived, if not actual, economic disadvantages associated with rural leadership.[44] The likelihood of such occurrences almost certainly increased when the household production responsibility system was adopted. The unattractive economic implications for leading cadres also appear to have been greatest for production-team cadres. Since they were already involved with physical farm labor, there presumably were fewer psychological barriers for team than brigade cadres to overcome if they decided to concentrate on the economic security and advantages of their households. The economic incentives to remain a responsible cadre in a brigade were not much better. Non-economic factors, however, came into play. On the assumption that brigade cadres could remain at their supervisory posts, their positions could continue to provide a measure of prestige. Thus, not all the former social and political amenities associated with rural and party leadership had been lost. At the very least, many brigade cadres were still able to avoid field work.

There were two areas in which the impact of the rural production responsibility systems on cadre income was unclear. The first involved communes and brigades where mixed production responsibility systems were used. The coexistence of different production responsibility systems within the same commune presumably resulted in quite different methods of calculating income for cadres who did the same kinds of work. This condition, in turn, apparently contributed to resentment, criticism of some forms of production responsibility, and even an unwillingness to serve as a leading cadre. The second area involved what *Hebei Daily* identified as problems for doctors, teachers, and technical personnel who were part of the system of collective social services. In particular, financial difficulties arose when local leaders had not determined "how to link such [welfare service] systems with the good or bad production of the production teams."[45] When the collective infrastructure was fundamentally retained, leading management cadres in some units had been negligent in establishing procedures to link the relatively set incomes of non-leading cadres to the economic successes or failures of their units. In still other units where mixed, quota, or household responsibility systems were adopted, some barefoot doctors were retained but were not paid. Sometimes doctors lost their positions through staff reductions, but no provisions were made to give them land or allow them to charge registration fees if they wished to continue practicing medicine.[46]

THE LEADERSHIP PROBLEM

The reform of the rural economy involved not only an indictment of longstanding economic policies and priorities, but also of rural leadership itself. Some leaders feared, therefore, that the reforms would involve political struggles. If leaders were again to be subjected to mass criticism, there were even fewer reasons to believe that being a cadre or leading party member could be "advantageous."

In early 1981, both *People's Daily* and *Guangming Daily* asserted that local leadership was the main problem in the countryside.[47] *Guangming Daily* observed, for instance, that "the poor management of the low-level cadres in some places and the poor performance of democratic management of financial matters and supervision by the masses of commune members" were a reflection of the general problem of poor leadership in the countryside.[48] Later in 1981, *Hebei Daily* pointed to the lack of "keen-witted and capable persons" among rural leadership groups, the low work efficiency of such groups, the lack of scientific and technological knowledge, and the low leadership and educational levels.[49] Such observations were not, of course, new revelations. Leadership in the rural areas had always been a problem.

In the context of the Third Plenum reforms, however, it was apparently feared that voicing such criticism too strongly would contribute to the waning morale of some rural cadres who might all too willingly "abandon their posts." Consequently, peasants and officials alike were warned repeatedly "not to make things bad" for local leaders; not to issue stern rebukes; not to make numerous replacements or engage in struggles or excessive criticism; not to revert to mass campaigns to solve problems; and not to act rashly in rectifying the economic, legal, political, or leadership mistakes of leading cadres. Even those cadres and peasants who had *objected to*, but had not obstructed, the adoption of rural responsibility systems were not to be criticized.[50]

CONCLUSION

By 1982, the national policy of "readjustment" was being applied to the rural as well as the urban, industrial economy. For some peasants this process portended a "drink of cold water in winter" and an end to the "policies of relaxation." Although high-level CCP leaders encouraged "flexible control," many cadres and peasants viewed this as implying greater "restraint," "restrictions," and retrenchment. To these individuals, the readjustment of rural policies represented a vindication of their warnings that chaos would ensue from "leading the cattle out of the communes" (that is, returning to individual, small-producer modes of production).[51]

The diversity of views regarding the adjustments in rural reforms underscores the extent to which those reforms differentially affected various groups in the Chinese countryside. It appears, for example, that peasants generally but not uniformly benefited from the efficiency generated by adoption of the rural production responsibility systems. In some instances, however, the breaking of the "iron rice bowl" created economic difficulties for rural families. Those farm families who did not prosper were relatively powerless to effect a shift back to egalitarian policies.

Questions of equity or efficiency, however, have not been the only concerns of the reform advocates. A strategically important problem has been the effects of the rural reforms on the organization and morale of grass-roots cadres. Rural cadres do not appear to have embraced the reforms with much vigor or enthusiasm. This is because the reforms, in particular the rural production responsibility systems, have complicated the supervisory tasks of rural cadres, eliminated the need for retaining some leading and non-leading cadres, threatened the economic security of many cadres, and lessened political leverage and prestige at the same time that greater political responsibility has been stressed for Party members. In short, the rural reforms of the Third Plenum contained virtually no benefits for cadres. The adoption of the rural production responsibility systems has apparently made it more difficult for the CCP to recruit and retain committed rural leaders in the face of competitive economic advantages.

CHAPTER SEVEN

Rural Collective Violence: The Fruits of Recent Reforms

ELIZABETH J. PERRY

To assess the impact of policy reforms is always a challenging task. And, when the reforms are as broad-ranging and multi-faceted as those implemented in China over the past half decade, assessment becomes especially difficult. Even if we take as our yardstick such seemingly straightforward economic measures as efficiency, productivity, or income, the conclusions are likely to prove complex and susceptible to a variety of interpretations. Still more controversial are political indicators of success or failure. How do we really determine whether government control has increased or declined? How do we ascertain whether mass interests are better or worse served by the new policies?

This chapter seeks to shed some light on the political implications of recent agricultural reforms through an examination of rural violence. In any society, the style and aims of collective violence are important indicators of the effects of government programs. Hence, an exploration of patterns of rural unrest in contemporary China should contribute to an assessment of the impact of the new policies in the countryside.

Traditional China, as is well known, lay claim to an exceptionally colorful history of peasant rebellion. Groups such as the Yellow Turbans, Red Eyebrows, and White Lotus contributed to an enduring and influential rebel tradition. China was also, of course, home to the largest—and perhaps most radical—agrarian revolution in human history. Millions of peasants joined the Communist-led Red Army to drive out the Guomindang and usher in a new socialist order. Yet, despite China's past experience of insurrection, when it comes to the post-1949 period most outside observers have pictured a quiescent peasantry. China's "blue ants" have been seen as completely subservient to official directives from above, the very embodiment of unquestioning obedience to authority.

In fact, however, the contemporary Chinese countryside has not been nearly so tranquil as popular Western images seem to suggest. Indeed, a surprisingly rich account of rural violence appears in the pages of the Chinese press itself. These sources reveal that violent protest against rural policies has erupted repeatedly in a variety of forms since the founding of the People's Republic.

EARLY YEARS OF THE PRC

Although we often speak of 1949 as the conclusion of the Chinese revolution, in fact, of course, Chinese society continued (or, in some places, began) to undergo revolutionary upheaval well after the consolidation of political victory in 1949. Carrying out this rural revolution involved far more than the mere pronouncement of new policy directives by top-level leaders. To implement these policies would mean inciting and then overcoming the resistance of unsympathetic and disenfranchised groups in the countryside.

During the first decade of the People's Republic, the Chinese press chronicled numerous cases in which disgruntled persons—

particularly those designated as landlords, rich peasants, or bad elements in the course of land reform—acted as leaders of protest movements to recapture their pre-liberation privileges. Frequently these movements evidenced the religious language and symbolism of sectarian groups in bygone days. But, whereas such appeals had traditionally attracted the bulk of their constituency from among impoverished or otherwise "marginal" elements (for example, peasants, hired laborers, women, peddlers), during the 1950s their participants were drawn disproportionately from the pre-liberation elite. Under these conditions, ironically enough, many of the sects developed a far stronger interest in land reform than had been true in the past. But now, of course, their plea was for a land reform that would restore the property of the dispossessed rural elite, rather than serve the interests of the poor peasantry. The groups were especially restive following periods of greatest government intervention in the countryside, most notably land reform and collectivization.[1]

The consolidation of collectivized agriculture in the late 1950s was, however, a watershed in changing styles of rural violence. After collectivization and the imposition of migration restrictions in 1958, villagers had little choice but to identify with the collective units to which they had been assigned. More important, there were powerful incentives to do so. Under the work-point system, everyone's best interests were served by cooperation among team members. Community began to replace class as the principal basis of collective action.

Ironically, the very success of state-sponsored collectivization in China gave rise to a strengthened local social organization. A study by William Parish and Martin Whyte argues convincingly that Chinese villages (that is, production brigades and teams) gained control of certain resources after the mid-1950s by processes of collectivization, land consolidation, and the like. Such trends served to enhance the probability that fellow villagers would act as a corporate body and to reduce the likelihood of communications among different villages. As Parish and Whyte put the matter, ". . . there have been shifts in social patterns in the countryside which in some ways make villages more encysted and closed to outside contact than they were twenty years ago."[2]

For rural cadres, whose incomes were also dependent upon their local unit, loyalty to kinsmen and neighbors sometimes threatened to outweigh adherence to state directives.[3] Especially in the area of production decisions, local leaders were often likely to promote methods that boosted the yields of their local units, even when such methods contradicted central policy. Nevertheless, the institution of collectivization—by placing control over the allocation of local resources in their hands—gave rural cadres a genuine stake in the system.[4] It was collectivization, after all, that finally replaced an old elite of landlords and rich peasants (whose economic power had, of course, already been undermined by land reform) with a new elite of rural cadres whose power rested in the disposition of collective resources.

Socialist transformation in the countryside served, therefore, to reinforce a collective consciousness among the peasantry. The logic of the system was such that cooperation with one's local unit was the *sine qua non* of successful survival in socialist China. Frequently, of course, the new rural units were synonymous with old patterns of identification: After 1962, communes were often coterminous with traditional marketing areas, brigades were often natural villages, and teams were sometimes lineages. But new socialist policies worked to strengthen horizontal identification within these local collectivities; land reform equalized property relations, collectivization leveled incomes, and migration restrictions bound peasants permanently to the same residential and work unit.

Thus, in the 1960s and 1970s, we already see an increased incidence of rural collective action based upon team, brigade, and commune membership. Reports of feuds between rival units[5] or of entire units withholding their labor from state-initiated projects[6] appear with some frequency in this period. However, only in more recent years—the period since the Third Plenum of December 1978, to be precise—has the official press begun to recognize rural violence as posing once again a serious problem in the countryside. And along with increased reports of local feuding and looting have come reports of a resurgence of "feudal superstition." Both phenomena, I suggest, are related to the reforms enunciated during the Third Plenum and its aftermath.

IMPACT OF THE REFORMS

Three aspects of the new reforms seem particularly important to an explanation of the recent violence. First, the implementation of agricultural responsibility systems has undermined the role of rural cadres. Once peasants are permitted to make their own decisions about farming methods and the disposition of agricultural surplus, there is presumably less need for administrative personnel to oversee these activities. Moreover, when payment is determined according to contract rather than by work points, cadres forfeit much of their control over the allocation of local resources. Restive cadres, we shall see, provide the leadership for most of the recently reported rural violence.

Second, the agricultural reforms stimulate conflict by redefining land and water rights. Although ownership remains vested in the collective (usually the production team), usufruct rights have been substantially altered with the introduction of household contracting and other agricultural responsibility systems. These alterations open the door to disputes over land boundaries,[7] water rights,[8] and woodlands[9]—the very stuff of which rural violence is traditionally made.

The combination of disgruntled local leaders and new pretexts for old feuds would seem to be the central force behind the resurgence of competitive violence in the countryside. But there is also a third feature which cannot be ignored: Many of the collective activities assume a religious demeanor. This phenomenon, it would seem, is related to the government's post-Cultural Revolution espousal of religious freedom. The recent relaxation of sanctions against officially approved "normal religion" has apparently encouraged a revival of unauthorized popular religious activities as well. On first glance, this particular feature of the movements would seem to resemble the sectarian protests of the 1950s. A closer look, however, makes clear that religious rituals have distinctly different meanings for participants nowadays. Rather than provide inspiration and identity to individual members of discredited social classes (as was the case in the sectarian uprisings of the 1950s), rural religious activities today serve to express kinship and community solidarity.

"Feudal Superstitious" Activities

Take, for example, a case of armed clan fighting which occurred in Changliu prefecture, Hainan, in the summer of 1981.[10] According to press reports, the hostilities had gained momentum in 1979 "with feudal superstitious activities of tracing ancestry, building ancestral temples and shrines, and public worship." A series of five armed clashes resulted in three deaths and scores of injuries. Houses were burned, and domestic animals and private property looted. In some production brigades, defensive walls were built around the villages, roadblocks were installed, and outposts set up. Weapons and grain were stored inside the fortified villages in preparation for the fierce struggles that lay ahead.

Before general warfare broke out, the Haikou Municipal Party Committee dispatched a work team of more than 600 state cadres, public security police, militiamen, and PLA soldiers. During a month of educational efforts, the work team was able to organize Party members of the feuding brigades into study classes. After considerable inner-Party criticism and ideological struggle, "most members of Party branches realized their mistakes, broke with feudal clan ideas, revealed inside stories of armed clan fighting and reached agreement on stopping armed strife and handing over guns."

The above case illustrates several common features of the recent violence: involvement of local Party members, collective organization, looting of material resources, and religious inspiration. Areas with long pre-liberation traditions of armed feuds (*xiedou*) seem especially vulnerable to the recent resurgence.

In the fall of 1981, Party members and cadres of a brigade in Dingan, Guangdong (another area known for its history of feuds), were charged with having planned, organized, and directed clan fights. The press reported that the feuds were associated with "superstitious" activities such as compiling genealogies, building and restoring ancestral tombs, offering sacrifices to the ancestors, and setting up clan associations. According to the analysis of the *Nanfang ribao*, "armed clan fights in the countryside had almost disappeared in the period before the mid-1960s." Due to the turmoil of the Cultural Revolution, however, local inhabitants had become "confused ideologically." Thus, during the past half year (that is, the latter part of 1981), armed clan feuds had again become "frequent occurrences" in some areas.[11]

Although the press is apt to locate the cause of rural violence in the realm of ideology, in fact one can usually identify a material basis to the disputes as well. Control over local resources is at the center of most of the reported conflicts. An illustrative case occurred in the spring of 1981 as members of a Hunan production brigade encouraged more than 2,000 members of the Zhang clan in four nearby communes to help exact revenge on a rival brigade. Under "the guise of paying respects to their ancestral tombs," the Zhangs were described in the Hunan press:

Waving banners, dancing the dragon dance, and with hoes and clubs in their hands, they wantonly sabotaged production; destroyed public property; trod on and uprooted over four *mou* of vegetables, rape, and wheat buds; cut down more than 200 large and small trees; destroyed enclosing walls and damaged commune members' houses.

The feud was evidently part of a longstanding dispute over graveyards which had burst into flame during the annual Qing-ming festival, the traditional time for visiting ancestral tombs.[12]

Inasmuch as freedom of religion is clearly guaranteed by Article 36 of the newly promulgated Constitution of the People's Republic, the authorities have found it necessary to differentiate between "normal religion" and "feudal superstition." An article in *People's Daily* on 15 March 1979 defined the latter category as referring generally to "such practices as witches, wizards, divine water, divine medicine, divinations by lot, fortunetelling, prayers for rain and children, curing diseases through exorcism, physiognomy and phrenology, geomancy, and other practices." At this time the official definition of superstitious activities did not include ancestor worship, and *People's Daily* explicitly cautioned against unwarranted application of administrative sanctions against such religious practices.[13]

Provincial authorities, on the other hand, were quick to stress the disruptive potential of community and clan worship. On 27 March 1979, a Zhejiang radio broadcast listed "repairing temples and building ancestral tombs" on a par with "seeking divine guidance" as deviations whose increasing appearance in the countryside would have to be dealt with harshly.[14] Later, the *Fujian ribao* also cited "building village temples" as a form of feudal superstition to be combatted just as firmly as "witchcraft, sorcery, use of elixirs, fortunetelling, astrological practices, invocation to avert calamities, rain-making, supplication for offspring, treating disease with exorcism, practice of physiognomy and geomancy."[15]

More recently, the central authorities have come to adopt a similar position. On 25 December 1982, the journal *Ban Yue Tan* carried an article entitled "Do Away with Feudal Superstitions," which was written by the *Ban yue tan* editorial department of the CCP Central Committee, the Ministry of Public Security, the United Front Work Department, the State Bureau of Religious Affairs, the Central Committee of the Communist Youth League, the All-China Women's Federation, and the State Administration of Industry and Commerce of the State Council. A 28 December Xinhua domestic broadcast called upon its listeners to "attach importance" to the *Ban yue tan* editorial, which noted with concern that some practitioners of "superstitions" had instigated "tens of thousands of people to participate in processions in which idols are carried, ancestor worshiping by all branches of a given clan, praying to god for magic cures, and other large-scale activities."[16]

The official distinction between religion and superstition is of some importance to practitioners, especially inasmuch as Article 99 of the new criminal law stipulates a fixed-term imprisonment of not less than five years for "those organizing and utilizing feudal superstitious beliefs, secret societies, or sects to carry out counterrevolutionary activities."[17] On 30 April 1981, the *Guangming ribao* carried a lengthy article intended to clarify the distinction, including this note: "Religion is different from feudal superstition in many aspects, but the most fundamental one is religion is a way of viewing the world while feudal superstition is a means by which some people practice fraud."[18] At the heart of the definition of fraudulent behavior are political criteria. The *Fujian ribao* put the issue quite bluntly: "Normal religious activities must be patriotic and law-abiding; support Communist leadership and the socialist system; have a legitimate organization recognized by the government departments concerned; be carried out within the law, and not interfere with politics, education, production, and social order."[19]

Despite official disapproval, however, "feudal superstition" has reportedly been gaining ground in the past several years. Some of the resurgence is almost certainly due to the recent relaxation of sanctions against "normal religion" and the difficulty many people apparently have in distinguishing between the two sorts of activity.[20] Other factors seem to be at work as well, however. One striking feature of recent reports from China is the frequency with which local cadres and Party members are portrayed as playing a leadership role in the revival of unauthorized popular religious activities.

Some of the incidents would seem, at least to an outside observer, fairly innocent. In the spring of 1980, for instance, the Party committee of Huaihua county, Hunan, issued a circular to criticize a Party member and cadre, Luo Runlan, for reviving the "four olds" in handling funerals. When her father died of illness in November 1979, Luo employed astrologers and Daoist priests to choose a location for the tomb and a date for the funeral. She also invited Daoist priests to say prayers for "leading the deceased."[21] When they arrived at the tomb all dressed in mourning clothes, Luo Rulan and her family members knelt down to worship. The county Party committee noted that this improper behavior had exerted "an extremely bad influence on the masses."[22]

Although Luo Runlan was evidently engaging in activities unbecoming to a Party member, for whom atheism is a requirement, it is not clear whether the same religious behaviors would have been deemed illegal had they been carried out by a non-Party member. Nevertheless, the official press has not limited its tales of cadre transgressions to these ambiguous cases. In May 1980, a more serious phenomenon was noted among "some minority Party members" (presumably Muslims) by the Qinghai provincial broadcasting service:

Some Party members and grass-roots cadres, won over by professional religious believers, still engage in and actively support religious activities.

> Some production teams stipulate that work points be given to commune members for their participation in religious activities and fines be imposed on those who do not participate. Some production teams use collective funds and materials to sponsor religious activities . . . Some production leaders lead commune members to hold ceremonious religious rites in which they themselves become mullahs or living buddhas. Some secretaries of party branches even ask mullahs and lamas for work instructions.[23]

The line between "religion" and "superstition" in such instances is difficult to draw. But, when the activities threaten to get out of hand, they are promptly designated "feudal superstition." Such was certainly the case in the spring of 1981 when the *Nanfang ribao* reported a "rather startling" news item. Several production brigades had recently organized a festival and parade under the name of the "great king and father touring the villages and driving away the evil spirits." A female commune member, claiming to be none other than the "great king and father," sat sedately in a sedan chair holding a long sword and dressed as a temple deity. She led a team of armed PLA soldiers who paraded in great style for several days:

> Even the schools suspended classes . . . Thousands of onlookers watched the proceedings, but nobody dared to come forth to stop the procession . . . What was even more startling was that those who carried the sedan chair and acted as vanguards of the parade were *all members of the Communist Party*.[24]

The *Nanfang ribao* report noted that, although by the mid-1950s most people had "untied the knot of feudal superstition," now:

> Many Party members and cadres . . . have become the captives and champions of these feudal superstitious activities. . . . Some people who formerly have led others in trying to break superstitious activities are now scared into submission by the sorcerers and witches and, of their own accord, have repented before the gods, seeking their pardon and promising to make amends. In this way, Party basic-level organs in the countryside have lost their fighting spirit against feudal superstitious activities.[25]

An article in *People's Daily* in the spring of 1982 described an incident in Zhangpu, Fujian, in which an old Party member had directed his own son and the son of the brigade Party secretary to "take advantage of the policy of religious freedom" by assuming religious leadership of the brigade. In this role, the young men succeeded in bilking fellow brigade members of some 2,900 yuan in "donations" to build a temple and seven Buddhist images and to stage two religious dramas. Each household was asked to pay an extra 40–50 yuan to provide cigarettes, tea, and liquor for those who came to see the performances. An assessment of two *jin* of grain per person was also levied for the personal use of the religious leaders. When the leaders threatened to collect an additional 5.7 yuan for yet another religious performance, some of the peasants balked. Threatened with having their doors nailed shut if they refused to comply, these brigade members wrote a letter of protest to the county people's government. An investigatory mission from the county established the facts of the case and returned the money spent for religious activities to the peasants.[26]

The above case suggests that, under some circumstances, expensive religious activities could actually drive a wedge between peasants and their local leaders. Although this case is probably quite unusual, it does indicate the active religious role often played by rural cadres and Party members. As the *Yangcheng Wanbao* reported on 20 May 1981:

Some cadres no longer go to the doctors when they are ill, but instead seek their cures through divine worship and the drinking of holy water. Moreover, some are even taking on the responsibility for temple reconstruction, collecting donations throughout the community.... Because of the active participation of Party members and cadres, the basic structure of agricultural organization has suffered a severe setback in the struggle to stamp out the practice of these feudal superstitions.[27]

Similarly, the *Nanfang ribao* reported in June 1981 that "Party members and cadres all participated" in the reconstruction of three temples in Gaoyao county. Interrupted by district Party members sent to disperse the gathering, one of the group leaders roused the crowd to an emotional frenzy. Thus incited, the mob destroyed the intruders' truck and beat up the Party delegates.[28]

Rural Cadre Unrest

The active involvement of local leaders in the recent religious revival would seem directly related to the increase in superfluous rural cadres—a problem recognized by the official press. *People's Daily* has complained that many commune and brigade cadres "think they have nothing to do now."[29] The *Ningxia ribao* described production-team leaders in its province as "in a state of paralysis."[30]

To deal with this situation, there have been calls for a reduction in the number of rural cadres. A Xinhua press release in August 1981 noted that too large a proportion of the peasants' produce was being reserved by collective units because too many administrative personnel were being provided with subsidies. To ameliorate the problem, it proposed that the number of brigade cadres be reduced to 5, while team cadres be reduced to 3.[31] At the same time, the press has argued that local cadres should continue to play an important leadership role. One listing of team cadres' functions included the preservation of public ownership of the basic means of production, curbing and correcting encroachments on public property, and mediation in civil disputes.[32] In short, rural cadres are enjoined to protect state interests in the countryside, even though their own bases of power have been substantially undercut by the new agricultural reforms.

The wishes of the state notwithstanding, there is ample evidence that many rural cadres have come to identify more and more with their local units. Even in cases where religious elements are entirely absent, rural cadre leadership is frequently behind the recently reported incidence of collective violence.

A case from the Huaibei portion of Anhui province gives an indication of the communal flavor of even the secular varieties of recent violence.[33] The niece of a Party member in a brigade in Fengtai county had had an altercation with a neighbor in her village. Distraught because of the manner in which the Party branch in her brigade handled the dispute, she took poison and died the following day. After learning of the death of his niece, Party member Chen Shijian gathered more than a hundred members of the Chen clan to wreak revenge on the neighbor with whom she had quarreled. The Chens marched to the village into which the niece had married and proceeded to stage an "eat-in" at the homes of the neighbor and her stepfather. They consumed all the available grain, pigs, and sheep, and made off with clothing and other items of daily use. When two cadres dispatched by the commune Party committee arrived to investigate, Chen Shijian led the attack against them. After the disturbance was finally quelled, Chen Shijian was expelled from the Party and his closest accomplices—Chen Shixian, Chen Shihao, Chen Shihou, and Chen Shichang—were sentenced to fifteen days' disciplinary detention for "fostering parochial prejudices of a patriarchal clan." Others who participated in the eat-in were ordered to pay 600 yuan in damages.[34]

An incident of large-scale looting, which took place in an orchard of Juancheng, Shandong, was also led by local cadres, a brigade Party secretary and a brigade militia commander. They directed a group of some 2,000 people who lived near the orchard to steal most of the apples belonging to one team in the orchard. Carrying jute bags, cloth sacks, and baskets, the intruders made off with 400,000 *jin* of apples during their two-day looting spree. They also beat up several orchard workers and stole carts, watches, cash, quilts, and so on.[35] Another very similar case of apple-grabbing was reported at a state farm in Wuhe, Anhui, later that month.[36]

Theft of state property was not an uncommon feature of these looting incidents. In January 1980, for example, commune members in the suburbs of Hefei, Anhui, made off with some 20,000 bricks from the provincial Academy of Agroscience. Tractor operators of two production brigades lent their machines to help transport the stolen goods. Investigation of the incident by the public security authorities established that a local production-team head had gotten into an argument with a member of the Academy of Agroscience. Afterwards, he and another production-team head (a woman), led members of their units to tear down the walls of the Animal Husbandry Institute.[37]

The pretext for a lucrative attack on state property could be remarkably slight. Wuchang Lake in Wangping, Anhui, opened for fishing on 19 January 1980. That day, Chang Xiaokai, secretary of the Party branch of the Dachang production brigade in Wuchang commune, got into a squabble with a policeman about a boy picking up a fish that had been dropped on the ground. One of Chang's nephews stood by him during the argument, while another of his nephews (also a member of the brigade Party branch) pulled an oar from a fishing boat in an attempt to strike the policeman. Seeing this dispute, fellow brigade members rushed to the police command post and beat up 16 leading cadres, public security cadres, and police who were trying to organize the fishing operations. Four state cadres were seriously wounded in the fray. Then Chang Xiaokai directed his followers to smash a truck of the Hefei aquatic products company and seven buildings being used by the command post. That afternoon, the local peasants looted 38,900 *jin* of fresh fish. Total losses to the state incurred in the day's riot were estimated at over 23,000 yuan.[38]

An *Anhui Daily* editorial attributed the above incidents to the "poisonous influence of anarchism spread by Lin Biao and the Gang of Four." These were interpreted as cases of "extreme individualism" in which interests of state and collective were ignored. A closer inspection suggests, however, that the actions were in fact quite organized; local Party, brigade, and team cadres served as leaders, while their kinsmen and units provided the bulk of the participants. Nevertheless, the final conclusion of the editorial was undeniable: "Some people feel free to gang up and steal grain, fruit, melons, fish, and so on, owned by the state or collective."[39] The prospect of real material gain was once again fueling competitive violence in the Anhui countryside.

In southeastern China, where armed feuds among rival clans had a long pre-liberation history,[40] these activities seem to have undergone something of a renaissance in recent years. For example, a fishing commune on Hainan Island experienced serious inter-brigade friction after some fights had broken out between children of the two brigades during the New Year's holiday. Instead of trying to calm the tensions, one brigade Party secretary convened a rally to further incite the peasants in his unit. With the support of the brigade militia commander,[41] he issued guns to his fellow brigade members and personally directed them in collecting stones and constructing fortifications. When peasants in the rival brigade saw these preparations, they reported the matter to the deputy secretary of their Party branch. However, "instead of taking the stand of Party principles, this man took a sectarian stand." When the two brigades had mobilized nearly 700 people, they opened fire on each other, aiming to kill. Six villagers were in fact slain, 5 injured, 60 houses destroyed, and large amounts of property burned or stolen. The total value of the damage was estimated at 47,000 yuan.[42]

CONCLUSION

According to the Chinese press, rural violence—of both the religious and secular varieties—has undergone a resurgence in recent years. Official explanations attribute this upsurge to incorrect ideology on the part of peasants and rural cadres. The dominant line of interpretation places the blame on the "calamitous years" of the Cultural Revolution, arguing that the policies of Lin Biao and the Gang of Four severely damaged the country's economy, culture, and education. Thrust into a state of dire poverty and ignorance by the misguided policies of that period, rural inhabitants began to fall back on familiar superstitions. By this analysis, the solution to the problem lies in "speeding up the progress of the four modernizations"; once China's material civilization has been developed, progress in "spiritual civilization will follow automatically."[43]

While scholars may debate the extent to which the 1966-1976 decade damaged the national economy, the political scars of the period are fairly clear. From the perspective of the current regime, one of the most serious costs is the fact that "quite a few cadres are still under the influence of leftist ideology."[44] As a result, many local leaders feel less than enthusiastic about post-Cultural Revolution proposals. The lack of commitment, in turn, means a decline in the state's capacity to enforce its will in the countryside.

Only occasionally do official discussions of the problem go beyond the expedient scapegoat of the Cultural Revolution to touch upon the impact of the current reforms. And, here again, the fundamental cause is seen as ideological. As a *Nanfang ribao* article put the matter, "Some cadres misunderstood the principle laid down by the Third Plenum of the Eleventh Central Committee concerning emancipating the mind. They believed that emancipation of the mind (*jiefang sixiang*) was tantamount to allowing wild and unbridled thinking..."[45] When a link to the Third Plenum is suggested, the solution is invariably seen as greater educational and propaganda work. The *Shaanxi ribao* noted in November 1980:

Since the spring festival, feudal superstition has been constantly practiced in certain parts of rural Shaanxi... These feudal superstitious activities have led to the fabrication and spread of many rumors... These rumors attack the Party's line, principles, and policies since the Third Plenum and slander and vilify the Party leadership and the socialist system... It is necessary to introduce scientific knowledge to the masses and vigorously conduct education in idealism and atheism. When their awareness is heightened... they will then spontaneously destroy superstition.[46]

Missing from such interpretations, of course, is any acknowledgment that the policy *content* of reforms enunciated at the Third Plenum may be responsible for inciting popular resistance; that such opposition may in fact be an expression of certain social *interests* disadvantaged or otherwise brought into play by the new agricultural measures. Instead, feudal superstition and rural violence are pictured merely as symptoms of confused ideology.

This is not to say, however, that the regime belittles the potential threat of such activities. To the contrary, the central leadership has indicated serious concerns for both economic and political reasons. On the economic side, the stated concern is that rural unrest will disrupt production.[47] On the political side, the worry is of counterrevolution. The authorities are especially fearful of the re-emergence of sectarian and secret-society activities. As a January 1982 *People's Daily* editorial put the issue, "Those who form reactionary secret societies which have been banned by formal decree will be punished according to law if the societies come to life again and resume activities."[48] More recently, the *Ban yue tan* notes:

In the past few years, there have been indications of a revival of reactionary superstitious sects and secret societies in some places. In some areas, scoundrels and counterrevolutionaries have appeared, claiming to be "emperors" or the "Jade Emperor descended to earth" . . . [49]

Similarly, an article in *People's Daily* of 12 July 1982 referred to "frequent reports" that in some areas people were taking advantage of feudal superstitious beliefs to call themselves emperor, make women their "imperial concubines," and disrupt social order.[50] In September 1983, Wang Hanbin, Vice-Chairman of the Legal Affairs Commission, identified seven types of criminal offenders for whom the death penalty should be more strictly enforced. One category consisted of persons who "organized reactionary sects and secret societies and used feudal superstitions to carry out counterrevolutionary activities."[51]

Just how widespread or serious such overtly political challenges are is impossible to determine. We are probably best advised not to exaggerate their significance. Rather, the revival of less dramatic forms of rural violence—based on kinship or community bonds and reinforced by popular religious practices— would seem to be a more pervasive phenomenon at present. The exact scale and frequency of even these more mundane varieties of competitive violence cannot be ascertained with accuracy. Their basic structure, social composition, and objectives appear, however, fairly clear. In contrast to the class-based, anti-state rebellions of the 1950s, today we see movements comprised of kinsmen or fellow villagers, led by rural cadres, and directed primarily against parochial targets for the purpose of material gain.

Instances of rural violence in contemporary China do not in themselves deny any positive contribution for the recent agricultural reforms, just as landlord resistance in the 1950s did not invalidate the benefits of land reform. An analysis of the violence that has occurred does, however, provide some insight into the social tensions generated by the new policies. Of particular interest at the present time is the prominent leadership role of local cadres. To the extent that such behavior represents a generalized problem of disaffection among rural leaders (as Richard Latham's chapter in this volume argues), it suggests that the agricultural reforms may prove more difficult to implement and sustain than the central leadership anticipates.

Future trends in the pattern of rural violence are, of course, extremely difficult to foresee. But if household responsibility systems remain in effect for some time, we may expect further changes. Most important, we should probably anticipate that collective action in future years will be organized less upon community and more upon kinship ties. With family replacing collective as the principal guarantor of income and welfare, blood ties should gain increased political salience. Furthermore, as income differentials within teams and brigades increase, community allegiances will probably decline. Rural families may once again develop a sense of class interest, with richer families preferring quite different policies from those favored by their less prosperous neighbors. The possibility of such divisions turning into open class struggle in the Chinese countryside remains extremely remote indeed. If, however, mobility restrictions were also to be lifted as part of the experimentation with market reforms, the future would be even less predictable.[52] For, were such a measure to be taken, rural China may begin to approach the scenario painted by analysts of peasant protest in Europe and the Third World: pressures of state and market acting in concert to undermine the village moral economy, thereby "freeing" peasants for participation in large-scale protest movements.[53]

Whether widespread rural violence will actually come to pose a serious problem for the Chinese authorities depends, of course, on government capacity as well as upon peasant and local cadre interests. To date, the speedy suppression of rural unrest (typically by the deployment of state cadres, security police, militia, and soldiers from county and commune levels) suggests that the state retains an impressive coercive capacity in the countryside. It remains to be seen whether reforms that directly affect commune and county cadres (for example, party rectification drives, establishment of township governments, county-level elections, and so on) will bolster or undermine the allegiance of state cadres in the rural areas. Should a substantial number of state cadres join the many local cadres who feel disaffected by recent policies, the outcome could prove disastrous from the perspective of the central authorities. But at present there is little evidence to suggest that such an alarming development is imminent.

PART TWO
Industry

CHAPTER EIGHT

The Politics of Industrial Reform

SUSAN L. SHIRK

In 1978, Deng Xiaoping initiated a set of policies designed to reform the Chinese economy. Like previous attempts at economic reform in the Soviet Union and Eastern Europe, the post-1978 Chinese reform initiatives generated political conflict. Predictions about the future of the economic reforms depend on our analysis of the source of this political conflict. Scholars and journalists observing China have identified the Communist Party as the major obstacle to economic reform. Although it is certainly true that many Party officials, especially the older veterans of the revolution, worry that economic reforms will make their political skills obsolete and diminish their power, the Party is not monolithic. The politics of economic reform are more complex than a simple "Party opposition" model would suggest. Party members and officials are differentiated along many lines, including geographic location and bureaucratic specialization. Party officials and government officials who manage the same economic sector are joined together in functional-bureaucratic "systems" (xitong)[1] or "complexes."[2] And Party and government officials who administer the same city or province work together as a local alliance to obtain more resources for their area.[3]

Economic reform in Communist systems like China becomes the focus of political conflict because it redistributes resources and responsibilities from some economic sectors, regions, and bureaucratic organizations to others. Partial reform of the economic system, leaving much of the structure, prices, and incentives of the centralized command economy intact, inevitably produces economic distortions and irrationalities. The groups who perceive that the reforms are detrimental to their interests then exploit these economic problems to argue against the reforms and force a return to centralization. After a period of time, when economic inefficiencies and popular clamor for improved living conditions offer vivid testimony to the shortcomings of over-centralization, academic economists and others favoring reform may get another chance. But piecemeal reform will probably once again end in re-centralization. In this way, economic and political forces interact to create a "cycle of reform."[4]

Will China become captive to a "cycle of reform" similar to that experienced by the Soviet Union and many of the countries of Eastern Europe? Or will it find a way to break out of this cycle to sustain the reform process? This chapter will attempt to answer these questions by analyzing the domestic political dimensions of the reforms in industrial planning, management, and trade. I seek to discover which groups have or have not benefited from the economic-reform policies; how these "winners" and "losers" have tried to influence China's post-1978 economic policies; and how the structure of the Chinese economic and political system has shaped this policy-making process. The chapter is based on interviews with economic officials in China conducted during visits in 1980, 1981, and 1982; interviews with Chinese economic officials visiting the United States;

interviews with emigrants in Hong Kong who had previously been employed in factories in the PRC; interviews with Western business people who have negotiated with Chinese economic officials; and analysis of articles from the Chinese press.

CHINA'S POLITICAL ECONOMY, 1949-1978

Before the initiation of the 1978 reforms, a set of industrial sectors, regions, and bureaucratic organizations was favored, protected, and subsidized by Chinese economic policies and structures. This set of groups, which could be called the "Communist coalition," consisted of heavy producer-goods industries (especially basic metals and machine building), the provinces of inland China, and the planning agencies and industrial ministries in the central government. The favoritism for heavy industry, inland provinces, and the central bureaucracy stemmed from the fundamental features of the Chinese economic and political system since 1949 (or at least since the beginning of the First Five-Year Plan in 1953). Despite periodic policy shifts, the Chinese system showed continuity in certain basic structural characteristics:[5]

(1) The Soviet-style *command economy*. Central planners set prices, wages, and output quotas, and allocated supplies and products according to national plan.

(2) The *centralized bureaucratic state*. Taxes and industrial profits were accumulated by the central government and reallocated as budgetary grants to the provinces and as investment grants to enterprises throughout the country. Industrial branch ministries oversaw all economic activity and had, in effect, national industrial monopolies.

(3) A *collectivized agricultural sector*. The central government invested few resources in the countryside. Most of the resources for agricultural investment and rural health and education were internally generated by peasants.

(4) The *extensive growth strategy*. A heavy industrial base was established and high growth rates were achieved by devoting a large proportion of national income to fixed capital investments in new factories, particularly in the producer goods industries.

(5) *Regional redistribution* through the central government. Centralized control over public finance and industrial investment was used, in part, to achieve greater regional equality by developing the inland provinces.

(6) The policy of *economic self-reliance*. China's door was closed to foreign investment. Foreign trade was held to low levels and aimed at the goal of import substitution. All foreign business dealings were conducted through the central foreign trade monopoly.

The cumulative effect of these structures and policies was to expand heavy industry rather than light industry or agriculture.[6] This expansion drive, which was unconstrained by calculations of the cost of capital, labor, or other inputs, created constant shortages of supplies. These shortages motivated enterprises to guarantee their supply of inputs through stockpiling and vertical integration and led provinces to strive for local self-sufficiency. As a result, levels of inter-regional trade were low.[7]

The Chinese system also created national economic monopolies, each headed by a ministry, as well as local monopolies for large provincial enterprises. Protected from the threat of either domestic or international competition, Chinese industrial enterprises and the ministries that administered them had little incentive to reduce their costs, raise productivity, or improve the quality of their products. Although centralized control over investment and planning was progressively weakened after 1957, and especially after 1966, central finance and investment policies were used to redistribute revenues from the more developed coastal provinces to the less developed ones in the interior of the country.[8]

Another effect of the economic structure was the centralization of power in the hands of the economic planning agencies, industrial ministries, and Communist Party organs in Beijing. The financial base of the state was built on the taxes and profits generated by industrial enterprises, and the industrial ministries—especially the heavy industrial ministries whose factories generated the lion's share of state revenues—became very powerful.

How has the loosening of the command economy, the shift in investment priorities, and the opening of China's door to international business, affected the set of group interests—heavy industry, inland provinces, and the central bureaucracies—that were favored before 1978? And how has the Communist coalition tried to stem the tide of reform? Before addressing these questions we must first examine the economic distortions and irrationalities that were produced when the reforms ran up against the unreformed features of the economic structure.

THE PROBLEMS OF PARTIAL REFORM

The first category of problems generated by partial reform of a command economy are those caused by irrational *prices*. There can be no efficiency gains from economic reform until the administered prices are changed to better reflect scarcity values or, even better, until prices are determined by market forces. This fact is well appreciated by Chinese economists. The redistributive implications of price reform would spark so much political conflict, however, that the Chinese leadership has decided not to attempt it yet. Meanwhile, the historical peculiarities of Chinese prices caused perverse results when the rules were changed to allow enterprises and local governments to retain a share of industrial profits. Government bureaus and factories rushed to enter the production of high-priced consumer goods; because raw materials were underpriced, processing industries, such as distilleries, cigarette factories, and sugar refineries, were the most lucrative. Even relatively backward inland provinces rushed to enter the consumer-goods market, erecting local blockades to protect their infant industries from competition with Shanghai or Tianjin brands. Since most manufactures continued to be distributed by plan and not by the factory itself, profit-seeking managers had more reason than ever to promote quantity instead of quality. In many factories, the quality of products declined under the reforms. Because commodities, fuel, and many producers' goods were underpriced, managers had no incentive to conserve them. Therefore, although output increased, there were no gains in enterprise efficiency, or what the Chinese call "economic results." Arbitrary prices also caused problems when China opened her door to foreign trade. Domestic and international prices were so out of line that the producers of some products had little interest in exporting (since their domestic prices were so high, while others (especially metals and other raw materials whose international prices were high) were so eager to export that there were domestic shortages. Perhaps most serious, under irrational price conditions, the profitability of a firm was an unreliable basis for evaluating the quality of its management or the productivity of its workers. A heavy industrial firm, no matter how astute its management or diligent its workers, would never make profits as high as even a poorly run factory producing televisions or some other high-priced consumer item. Any factory that consistently operated at a loss could defend itself against the threat of being closed down or merged with another firm by claiming that its losses had external (price) causes—not internal (management) ones.

A second set of problems stemmed from the *expansion drive,* a fundamental feature of the command economy which was intensified by its partial reform. In the past, Chinese managers were motivated to expand the capacity of their enterprises because investment capital was given to them as a grant, and managers were evaluated solely on the basis of physical output. They never had to worry about selling their products or about the costs of production. A new workshop would make it easier to overfulfill the output target, and the costs of construction, additional machinery, and labor would simply be borne by the industrial ministry. The new profit incentives heightened managers' eagerness to expand their capacity and their output, and gave them the wherewithal to expand. The fiscal decentralization turned local governments into economic entrepreneurs, ready and able to create new factories—especially in the lucrative processing industries. The banks, which for the first time were given the right to make loans, added fuel to this industrial expansion. Local bankers were eager to help local officials expand into profitable consumer-goods markets. The ability of enterprise managers to spend some of their profits and of local officials to spend some of their profits and taxes, made it much more difficult than in the past for central authorities to control the level of investment. Despite the center's efforts to cut investment, it increased, especially at local levels, during 1982 and 1983. When the central authorities tried to persuade localities and enterprises to concentrate on technical upgrading of their equipment instead of building new factories or workshops, they had little success. The funds provided for technical transformation were, instead, often used for "expanding the production capacity of ordinary processing industries."[9] As a result of the investment explosion in local industry, the central government found itself without the funds for large transportation and energy projects. Industrial output and revenues had increased, but the center was getting a smaller and smaller share of them.

A third problem was the *supply shortages* created by the expansion of industry. Chinese managers have always sought to stockpile their inputs because, in a centrally planned economy, one can never be sure of obtaining materials and equipment that meet specifications and are on time. Shortages were then worsened by the vast amount of supplies that lay idle in factory warehouses. The expansion drive accelerated by the 1978 reforms increased the demand for construction materials and machinery and raised the incentive for managers and local officials to hoard them. Shortages were exacerbated by the fact that most of such basic materials as concrete, steel, and glass were no longer distributed by central plan (as a result of the expansion of local plants producing these materials during the Cultural Revolution decade). At present, only 27 percent of total output of concrete and less than 50 percent of plate glass, is distributed by the national planners.[10]

Local authorities are reluctant to permit these vital materials to be sold to other areas for fear that shortages may sabotage their efforts to build up local industry. Despite the intention of central policy-makers to shift investment priorities from heavy to light industry, the expansion of local light industry paradoxically stimulated the growth of such producer-goods industries as steel, coal, and machinery. This expansion of heavy industry exacerbated shortages of raw materials and power.

A fourth problem created by partial reform lay in the realm of *labor costs*. Managers in socialist command economies have always sought to expand their labor force and increase wages; labor, like other inputs, was stockpiled so that enterprises could meet their production quotas with greater ease. In the past, however, Chinese managers had no power to give raises or bonuses on their own initiative. Wage increases were infrequent and were carried out on a national scale, and bonuses were determined by the industrial bureaucracies as a set percentage of the wage bill. The 1978 industrial reforms not only gave managers a new profit motive to hire more workers and ensure their cooperation through pay raises, but also gave them the right to distribute bonuses out of their profits for the first time. At the same time, the national leaders approved three nationwide wage increases and encouraged the use of piecework pay and bonuses in the hope of raising labor productivity. Chinese managers took advantage of this new situation to do whatever they could to satisfy their workers. They concentrated most of their retained profits on new housing and bonuses for their employees. Because distribution of bonuses according to work performance was socially divisive and resisted by many workers, managers allowed bonuses to be distributed in a roughly egalitarian fashion.[11]

Because they still lacked the power to increase wages within their own factories, some managers resorted to illegal methods—such as giving away free or cut-rate consumer goods (produced by their own plant or obtained by various types of covert inter-enterprise exchanges)—to reward their employees. As a way of stockpiling labor and creating jobs for their workers' children and relatives, many enterprises set up large collective factories. There were also frequent complaints in the press about managers who had used illegal methods to hire their employees' relatives out of rural villages.

It would be wrong to interpret these actions by management as an indication of the bargaining power of Chinese workers.[12] Unlike Eastern Europe and the Soviet Union, China has no labor market as yet, and also has a surplus of industrial labor. Managers do not have to worry about losing their workers and not being able to find replacements. It is therefore something of a mystery why Chinese managers are nevertheless so anxious to keep their workers happy (the answer awaits further research on the career incentives of Chinese managers). In any case, the actions of Chinese managers since the implementation of the 1978 reforms have clearly resulted in a massive increase in the cost of labor, without producing significant improvements in labor productivity.

A fifth problem produced by the partially reformed post-1978 Chinese economy was *inflation*. The increases in spending by local politicians and enterprise managers sparked by the recent reforms may not have led to improvements in enterprise efficiency or productivity, but they did lead to higher prices. The repressed inflation (created by the shortages which are a constant feature of Soviet-style centrally planned economies) of the past was replaced by more overt inflation. The reforms permitting free markets in food and sundries in both cities and countryside and raising the pay of workers and peasants pushed prices upward. The most serious inflationary pressures came, however, not from individual consumers, but from local politicians and factory managers who bid up prices of producers goods as they rushed to expand their local industrial bases. Because central planners had lost control over a significant proportion of such basic items as concrete and steel, local suppliers raised their prices in response to the increase in demand for their products. The border prices (the actual prices of sales across provincial lines) of construction materials and equipment rose even higher than the local prices. During the summer of 1983, the central leaders pleaded with local suppliers to lower their prices to within range of official levels and urged local political entrepreneurs to delay or cancel local construction projects, but it will be difficult to reassert control over local producer-goods markets.[13]

The problems highlighted by the post-1978 Chinese economic policies—price irrationalities, the drive to build new factory workshops instead of upgrading existing ones, shortages of supplies, overpayment of workers, and inflation—suggest that, when a Soviet-style command economy undergoes partial reform, its systemic weaknesses are exacerbated.

Partial reform also intensifies competition among localities, bureaucratic organizations, and enterprises. The decentralization of profit-retention and decision-making power gives local political units and enterprises both the incentive and the capacity to promote their organizational business interests. They expand their industrial empires with entrepreneurial zeal because the remaining structures of the command economy protect them from risk. Because substantial profits and tax revenues are now at stake, various levels and organizations of government have strengthened their sense of "ownership" over enterprises, and competition between units has intensified. The economic competition between government units tends to increase the balkanization of markets in manufactures as each local government seeks to use blockades and other "administrative methods" to protect its own factories from outside competition. Only in foreign trade and investment is there true nationwide competition, with each city or province seeking to attract investment or make export sales by offering lower prices and better goods to foreign businesses. In domestic business, however, there is little "real" economic competition among provinces or cities. Local politicians concentrate instead on retaining "ownership" of their enterprises, protecting local markets, and negotiating with the upper levels of government for good rates of profit and foreign-exchange retention. Regulations about what share of profits and foreign-exchange earnings can be retained by local governments are so flexible that there is constant bargaining between enterprises and local governments and between local governments and the center.

The economic reforms of 1978 have, then, not only had a differential effect on various sectors, regions, and levels of government, but they have also altered the political-economic context within which these groups operate. Economic competition among units has increased, but competition is still waged with the techniques of bureaucratic warfare rather than market competition. And the groups that have the greatest economic advantages under the recent reforms—light industry, coastal provinces, and local governments—often find themselves outmaneuvered in these struggles by the members of the Communist coalition—heavy industry, inland provinces, and the central ministries—which have superior bureaucratic clout.

SECTORAL CONFLICTS

What has been the effect of the recent reforms on heavy and light industries in China and how will this sectoral conflict influence the future of economic reform? Light industry undoubtedly has benefited more from the reforms than heavy industry. For one thing, light industry has greater flexibility than heavy industry and therefore could respond more quickly to the growth in consumer demand from workers and peasants now with more money to spend. (This flexibility stems from the fact that, even in a command economy, light industry tends to be more "consumer-oriented" than heavy industry; that it is more often locally controlled and organized as a collective; that the prices of its products are less tightly controlled by the center; that it is more labor-intensive than capital-intensive; and that its workers, more of whom are female, have lower expectations.) Light industry also has wider opportunities to enter international markets. Foreign importers and investors are much more interested in Chinese textiles and handicrafts than in Chinese steel or machinery, which are low in quality and technological sophistication. The shift in investment priorities at the center obviously worked to the advantage of light industry. And the new policies allowing enterprises to retain a share of their profits also favored light industrial factories which, as a rule, are more profitable because their products have been assigned higher prices. Although even light industrial plants faced some difficulties adjusting to the reforms—for example, because they still were responsible for marketing only a small share of their own products, they were slow to adapt them to consumer preferences, and the massive expansion of local factories created intense competition for supplies and sales—they were certainly better situated to take advantage than was heavy industry.

Heavy industrial plants, on the other hand, found their capital investments and output quotas cut by the center's readjustment policies. Although they maintained their traditional claim for priority in the distribution of scarce electric power and railroad freight transport, competition for power and transport was increased by the expansion of local processing plants. The new profit-retention policies did heavy industries little good, because, no matter how well their factories were managed or how hard their workers worked, irrational prices meant that they would always earn lower profits than, say, a tape-recorder or a cigarette factory. The new open door offered heavy industries few new opportunities. Some companies are attracted by China's reserves of coal, petroleum, and nonferrous metals, and have been willing to invest in projects to develop and export them, but iron and steel and machinery plants find that their antiquated equipment, erratic product quality, remoteness from ports, and lack of familiarity with the international market frustrate their trading ambitions.

Heavy industries, moreover, feel the threat of international competition most keenly. The machine-building industry was a major beneficiary of the extensive growth and self-sufficiency policies of the past. It supplied equipment for almost all Chinese factories under monopolistic conditions, and expanded at approximately 20 percent a year during the 1950s-1970s.[14] By 1978, there were over 100,000 machinery-manufacturing enterprises in China, nearly one-third of the total number of industrial enterprises (because of the expansion of processing plants in recent years, the ratio is now closer to one-fourth).[15] The Ministry of Machine-Building (MMB) administers 11,000 of these enterprises, which employ 5.5 million employees.

When the new open-door policy made it possible for Chinese factories to modernize with purchases of imported equipment, the machine-building industry fought to maintain its monopoly and protect its domestic market from international competition. It was successful in re-establishing a 1950s institution, a special Equipment Approval Division within the MMB. This division is empowered to approve all factory imports of equipment, even those produced by foreign-Chinese joint ventures in China.

A request for imported equipment is submitted to the State Planning Commission, which in turn sends it to the division. If the Ministry determines that one of its factories can produce the same piece of equipment (regardless of cost), it vetoes the import. Foreign business people report that the Ministry of Machine Building has in a few instances even prevented the Metallurgical and Petroleum Ministries from sending representatives abroad to shop. Several disputes over equipment imports between the Machine Building Ministry and other industrial ministries have had to be resolved by the State Economic Commission or State Planning Commission. Machinery industry protectionism has clearly resulted in a decrease in purchases of foreign machinery: Imports in the first half of 1982 declined 43 percent from the same period in 1981.[16]

The opposition of heavy industrial systems (*xitong*) to economic reform goes beyond the issue of protectionism. There are numerous indications that, when the economic reform measures were first implemented in 1978-1980, the political officials who represent the interests of heavy industry objected to them. The so-called "petroleum faction" advocated a continuation of centralization and the extensive growth strategy with heavy industry maintaining its preeminence; the only major departure they favored was the financing of industrial modernization with foreign investments and petroleum exports. As Dorothy Solinger has shown, the delegates from provinces in which heavy industry is concentrated were the most vocal critics of reform at the Fifth National People's Congress; they were particularly opposed to the readjustment of investment priorities (a reduction in central capital investment and a marginal shift in the capital allocations from heavy industry to light industry and agriculture).[17] Managers at heavy industrial factories whom I interviewed in 1980, while not expressing outright opposition to the reforms, said that they were of only minor benefit to their operations.

The opposition of heavy industry constitutes a serious threat to the future of the economic reforms, because heavy industrial systems exercise considerable influence over policy decisions in a socialist political economy. This political influence stems, first of all, from the structure of the command economy. All economic sectors (such as the machinery industry, the petroleum industry, the iron-and-steel industry, and the light consumer-goods industry) are organized into vertical national bureaucracies headed by ministries (called "branch ministries" in the Soviet Union and Eastern Europe) in the capital. Each minister, sitting in Beijing, is able to articulate the interests of the industry he represents. In this way, the structure of the socialist political-economic system gives industrial sectors, and heavy industry in particular, a powerful voice in policy-making.

The preeminence of the heavy industrial ministries also stems from the economic planning process in a command economy. Planning by material balances means that the sectors who produce equipment or materials for many other parts of the economy are in a pivotal position. The scale of the planning process forces officials, who have too many items to handle, to identify some products as priorities.[18] Because so much of the economy needs coal, steel, and machinery to fulfill the plan, these industries are in a strong position when the bargaining over plan targets and investment allocations occurs. Their status as crucial industries also gives coal, metallurgy, and machine-building bureaus at the provincial and local levels a strong claim on resources such as electricity and manpower.

The political strength of heavy industry can also be explained by the extensive growth strategy which the Chinese adopted from the Soviet Union when it initiated its First Five-Year Plan in 1953. Chinese economic officials I have interviewed point out that, when the central economic bureaucracy was established after 1953, the most able cadres from the provinces were recruited into the heavy-industrial ministries because of the priority on heavy industry stated in the Plan. From that time onward, it has been widely recognized that the leadership of the heavy-industrial bureaucracies is of superior caliber. The preeminence of heavy industry is also reflected in Chinese national wage scales. The workers in heavy industry are paid on a higher scale than the workers in light industry; and the managers of major heavy-industrial plants, such as the Anshan and Wuhan Iron and Steel Companies, have a cadre rank higher than some provincial governors.

The political influence of the heavy-industrial systems can also be explained by the fact that their taxes and profits constitute the lion's share of the central government's fiscal base. The government's financial dependence on heavy industry means that, when the minister of coal, metallurgy, or machine-building asserts himself in a policy issue, he is likely to prevail over ministries like education or health, which contribute no revenues, or even like light industry or agriculture, which contribute fewer revenues.[19]

For the most part, heavy industry seems to have used its bureaucratic power to get the best deal it could in the new environment created by the 1978 economic reforms, not to make a direct challenge to the reforms themselves. The heavy industrial ministries have maintained central control of more of their own factories than have the light industrial ministries, which have lost control of most of their factories to local authorities. The heavy industrial ministries have successfully challenged the monopoly of the Ministry of Foreign Economic Relations and Trade (MOFERT) to establish their own trading companies, while the Ministries of Light Industry and Textiles have (with a few exceptions) been unable to do so.

The heavy industrial ministries usually come out on top in inter-agency conflicts. With the advent of the reform era, conflicts between ministries have become more frequent and often have required higher-level leaders to intervene. For example, the Ministry of Machine Building was eager to enter the lucrative consumer-goods market with washing machines and refrigerators, but the Ministry of Light Industry resisted this challenge to its monopoly. The State Planning Commission called a meeting among several ministries to divide the burgeoning market in ten high-volume consumer products. The Ministry of Machine Building emerged victorious, with a sizable share of the washing-machine and refrigerator business (in washing machines, it was awarded the multi-cycle and washer-dryer machines, while the Ministry of Light Industry kept the simple machines). At the same time as it won a share of the consumer durables market, the Ministry of Machine Building found itself with a huge new market in mini-tractors and other small farm machinery. Individual peasant households, now assigned their own fields under the rural responsibility system, have been spending more money on farm equipment and less on consumer goods than expected. In addition, the rapid expansion of local processing facilities created a demand for new machinery, which the machine-building industry has been surprisingly quick to supply. These new sales opportunities have helped the machinery industry rebound from the shocks of reform and readjustment.

The representatives of heavy industry certainly appear to have had a hand in the effort to re-centralize industrial investment since 1982. The political line on economic reform during 1982-1983 has been that its most serious problem is the loss of central control over investment capital. With so much revenue being retained and reinvested by local governments, the center lacks the resources to build the major energy and transportation projects China needs to modernize. The re-emphasis on centrally funded large construction projects naturally appeals to the heavy industrial sector. It means more investment in heavy industry from the center and a steady demand for coal, steel, and large machinery. Although the officials of heavy industry may have been surprisingly adept at responding to the changes in demand created by the post-1978 reforms, like industries in the United States which supply most of their products as contractors to the federal government, they prefer the predictability of selling to one large bureaucracy to the risks and uncertainties of a more diversified market.

REGIONAL CONFLICTS

How have the recent economic reforms affected the relationship between inland and coastal regions in China and how will regional conflicts influence the future of economic reform?

As a result of the reforms, the coastal provinces (Liaoning, Hebei, Shandong, Jiangsu, Zhejiang, Fujian, Guangdong, Tianjin, Beijing, and Shanghai) appear to be regaining the position of economic superiority they had held before the Communists came to power in 1949. Before the Communist era, China had a dual economy. Foreign trade and modern industrial production were concentrated in the coastal region, which had experienced economic and cultural penetration by foreigners. The inland provinces had traditional agricultural economies and little industry except for such industrial enclaves as Hankow—now part of Wuhan—and Chongqing.

Although there were periodic shifts in policy, generally speaking the policy of the Communist leadership from 1956 until 1978 was to disperse China's industries throughout the country. The rationale was military defense and social equality, but the logic of the decision was political. The Communist movement had met significant opposition in the more cosmopolitan, urbanized "white areas" of coastal China during the 1920s and 1930s; it had found the popular support that was critical for its victory in the poor, agricultural regions of inland China. After coming to power, the Communists favored the inland areas with a disproportionate share of central capital investment and fiscal subsidies.[20]

Cities like Shanghai had their industrial revenues siphoned off by the central government and redistributed to the inland provinces. (Some of this redistribution was simply to compensate for the regional inequalities caused by the irrationally administered price system. Shanghai's industrial profits were inflated by the low price of cotton and the high price of finished textiles, while inland cotton-growing provinces were in the opposite situation.) Because the goal was to create complete economies in the inland provinces, these funds were used to build steel mills, automobile and tractor factories, and other heavy industrial plants in every province. Despite these policies of regional redistribution, the coastal areas continued to grow at rates at least equal to the interior.[21] The coastal areas continued to prosper because, as Audrey Donnithorne has shown, after 1957 central control over industrial investment became progressively weaker, and, despite its discriminatory treatment by the center, Shanghai and other coastal cities continued to benefit from the modern industrial plant and skilled manpower built up before 1949.[22]

An essential component of the 1978 reforms was the freeing up of the coastal provinces from central controls to take advantage of their natural strengths, not only their industrial plant and skilled manpower, but also their port facilities and ties to Overseas Chinese capitalists who prefer to invest in their home provinces. As one newspaper article explained, "Coastal cities have good economic foundations, developed transportation and communications systems, and a long history of establishing economic and technical contacts with foreign countries. They should have more decision-making power to promote foreign economic relations. We should give full play to their advantages."[23]

The opening of China to international business has clearly benefited the coastal more than the inland areas. The new foreign trade and investment opportunities, combined with the policies allowing local governments and enterprises to negotiate independently and retain a portion of their foreign exchange earnings, have stimulated economic competition among Chinese cities and provinces. Local authorities seek to develop their local economies with the profits of international commerce in a manner reminiscent of the local authorities in the mid-nineteenth century who sought to build local armies with the likin taxes collected along domestic trade routes. In this new competitive environment, the coastal provinces appear to be winning most of the prizes. Foreign investments are concentrated in a few coastal provinces and municipalities, namely Shanghai, Tianjin, Dalian, and the Special Economic Zones (areas granted special powers to offer concessionary terms to foreign investors) in Fujian and Guangdong provinces. The four Special Economic Zones (SEZs) were able to garner 60 percent of direct foreign investment in China in 1981.[24] As centers of light industry, the coastal areas are also the source of a large percentage of China's manufactured exports. For example, 1,700 of Shanghai's 8,000 factories are now engaged in producing for export.[25] The coastal ports also ship exports and imports for many inland enterprises; despite increasing competition from inland ports up the Yangtze, the port of Shanghai still handles one-fifth of total national exports.[26]

The new open-door policy has created a bonanza for the coastal provinces, especially Guangdong. During 1981, Guangdong reportedly earned over U.S. $1 billion in foreign exchange.[27] Foreign-exchange earnings can be used to import foreign equipment and materials; and, despite the formal prohibitions against domestic sales of foreign exchange, local officials sell foreign currency on the black market to obtain Chinese currency for construction projects. Because coastal areas are attractive sites for foreign investment, potential investors can sometimes be persuaded to finance local energy projects and road construction.[28]

The center is not only allowing the coastal provinces to keep a bigger share of their business earnings but is also enhancing their attractiveness to foreign investors by concentrating infrastructure investments in these areas. In theory, through building on the stronger international position of the coastal regions, the development of the whole national economy will benefit.[29]

While the regional implications of the new open-door policy are clear, the implications of other measures in the 1978 reform package are more complex. As producers primarily of grain rather than cash crops, inland farmers gained less advantage from the liberalization of rural marketing and the agricultural responsibility system than did the farmers on the coast. Nevertheless, the inland provinces of Anhui and Sichuan were the pacesetters for agricultural decentralization. Because the economies of the inland provinces depend more on heavy industry than light industry, the shift in investment priorities threw them into a slump. Even so, the inland cities of Chongqing and Wuhan, which have strong local economies based on heavy industry, seem to have been enthusiastic about the reforms giving local firms and governments more autonomy.

Inland local politicans used this autonomy to try to break into the more profitable light consumer-goods market. When their infant consumer-goods industries had a hard time competing with the brand-name bicycles, watches, and televisions from Shanghai and Tianjin, they pleaded for official protection from the center or established local blockades. Although the center often condemns such local trade restrictions as "administrative interference," it sometimes acquiesces. For example, in 1982 Anhui province objected to an exhibit of Shanghai products held in their province because they were themselves now producing many of the same items in local factories. The State Economic Commission refused to grant Anhui's request that the exhibition be closed down, but it did forbid Shanghai to display articles identical to those made in Anhui.

The complexity of the regional implications of the economic reforms is highlighted in the case of Shanghai. This coastal city, which has traditionally been China's center of industry, trade, and commerce, was until 1983 kept on a tight leash by the center. When Guangdong and Fujian were given the freedom to deal directly with foreign businesses and offer them concessionary terms, Shanghai was not. When the local governments were given the right to retain a share of their revenues and foreign-exchange earnings, Shanghai, along with Beijing and Tianjin, was not given similar financial autonomy. The rationale for maintaining the fiscal restrictions on these three was that the central government could not afford to give up the revenues from the three municipalities, which amounted to about 25 percent of the national total.[30] Economic officials I have interviewed, however, believe that the restrictions also reflect the central leaders' anxiety about a challenge from Shanghai. (Whether they fear Shanghai's Cultural Revolution radicalism or its economic power is not clear.)[31]

The net effect of the entire set of reforms on the inland provinces and their political response to the reforms are difficult to assess. There are some indications that many officials from inland provinces believe the reforms put them at a competitive disadvantage and that they have advocated re-centralization to protect their economic interests. Noted Chinese economist Xue Muqiao observed that there were serious contradictions among regions caused by "extremely uneven industrial development": "In general, the industrially developed areas wish to acquire greater independence and the underdeveloped ones prefer unified management and unified allocation of products by the central government. For these reasons, it has been very difficult for the state organs of economic management to reach an agreement on changing the current system of planning and management."[32]

Some officials from the inland areas may have tried to sabotage the economic reforms—the new open-door policy in particular—by publicizing the danger of corrosion of Chinese culture by decadent ideas and lifestyles from abroad. The inland areas have always been more culturally parochial than the coast. Many of the scare stories about the infiltration of bourgeois foreign culture come from the inland provinces. Complaints about pornographic pictures and tapes (called "yellow materials") imported from abroad have been heard from areas as deep in the interior and remote from foreign contact as Shanxi.[33]

Some officials from the coastal provinces also believe that recent campaigns against corruption—which have focused on cases of smuggling and shady business practices along the coast, in Guangdong in particular—reflect inland resentment of increasing coastal prosperity. Coastal officials were forced to defend their integrity and the national benefits of the open-door policy at the National People's Congress in 1982.[34] Despite the criticisms of the open-door policy, which may be motivated in part by the jealousy of inland officials, the door has remained open. Officials from the Special Economic Zones sometimes complain of administrative restrictions from Beijing, but generally they feel that the Special Economic Zone Office, which is directly under the State Council at the apex of the governmental structure, has provided effective representation of their interests.

Although the regional implications of the economic reforms are difficult to sort out, the reforms undoubtedly have worsened inter-regional strains. The Chinese press frequently reiterates that the reforms are in the interests of the entire country even if they cause some regions to "get rich first."

The political problems of regional unity and equity created by the reforms have put considerable strain on the central leadership and have caused them to consider new forms of national economic integration. In the past, regions were held together by their dependence on the center. The planners in Beijing determined all allocations on a national basis, and, except for some barter trade, there was little direct economic contact among provinces. Because the center has already lost much of its control over local spending and allocation of supplies, it is now allowing governments and enterprises to have direct economic transactions with one another—including joint ventures as well as trade. This "inter-provincial cooperation" is viewed by the leaders in Beijing as a way to ease inland shortages of capital and technology and, possibly, as a "side payment" in the reform package to disarm inland opposition to it.[35]

Inter-provincial cooperation also represents the beginnings of capital transfers in China. For the first time, a coastal enterprise or industrial bureau can consider alternative uses of its retained profits. The coastal units naturally see an investment in an inland mine or factory as a good way of guaranteeing their supplies of raw materials, energy, and other inputs. It is not yet clear whether coastal units are investing primarily in the extraction of raw materials or are transferring some of their production operations into the interior; and evaluation of the benefits of inter-provincial cooperation for the inland provinces awaits such data.

There is also uncertainty about the central bureaucracy's attitude to direct economic relations among provinces. Although inter-provincial cooperation has been encouraged by the official press as part of the reform package, the central planners worry about losing control. They complain that inter-regional transfers of commodities arranged by the units themselves have pre-empted shipments required by the national plan. Sometimes the planned shipments are removed from trains and left to rot on the platform; they are then replaced in the freight cars by the inter-regional trade in which local authorities have a more direct financial stake. To prevent this type of interference with the national plan a special office to regulate inter-provincial cooperation has recently been established in the State Planning Commission.

The strains on regional relations produced by the economic reforms raise important questions about the way regional interests fit into the national policy process. Industrial sectors are represented by their ministries, but how do regional interests come into play? There are representative offices from every province located in Beijing, but we do not know what role they play in the policy process. Are certain regions more heavily represented in the personnel of certain ministries? For example, are many officials in the Ministry of Machine Building from the northeast where many of the largest factories are located? If so, do they constitute regional factions with ties to top level political leaders? (It has been noted that a number of officials with ties to the south, especially Guangdong, have risen to political power during this era of economic reform.)[36] Each vice-premier is assigned responsibility for several ministries and serves as a kind of patron to these ministries. What role do these politician-patrons play when inter-agency and inter-regional conflicts are resolved by the Standing Committee of the State Council? Is this where bureaucratic and factional politics intersect? The fate of economic reform—as well as decisions about the location of major infrastructure projects such as nuclear plants—will be strongly affected by the way sectoral and regional interests interconnect in the policy process.

BUREAUCRATIC CONFLICTS

The new opportunities and risks created by the reforms have intensified the conflicts among bureaucratic agencies in Beijing. Articles in the press criticize agencies for thinking only of their own interests and for engaging in constant "wrangling in economic work" and "arguing back and forth."[37] For example, the Petroleum and Geology Ministries have fought over control of offshore oil exploration, which is being carried out with foreign participation. The Ministry of Finance has clashed with these two ministries and other industrial ministries and localities over the tax rates for joint foreign-Chinese ventures. The Ministry of Finance wants to take in more taxes, while the industrial officials—who care more about foreign-exchange earnings and technology transfer—offer concessionary tax rates to attract investors.

The net effect of the reform package on the relative power of the industrial (branch) ministries and the central planners and regulators (the State Economic Commission, the State Planning Commission, the Ministry of Finance, and the Ministry of Foreign Economic Relations and Trade) is not yet clear. The decentralization of the fiscal system to local enterprises and governments certainly created great uncertainty for the Ministry of Finance. The Ministry, however, has recently prevailed in a major policy decision to "replace profits with taxes" (*li gai shui*). Enterprises will have to pay a 55 percent tax to the central treasury before retaining a percentage of the remaining profits, thereby assuring the Ministry of Finance a steady flow of revenues. The State Economic Commission (SEC) and State Planning Commission (SPC) also lost much of their control over economic activity owing to the reforms (as well as the decentralization of supplies and construction which preceded them). The role of the SEC and SPC as economic arbiters has increased, however, as they resolve the conflicts among ministries and localities created by the new environment.[38]

The Ministry of Foreign Economic Relations and Trade appears to have lost power and status despite the increase in international business activity. Its monopoly over foreign trade and investment was destroyed by local authorities and industrial ministries who established their own trading companies. Even though MOFERT Minister Chen Muhua pleaded for a recentralization of foreign trade and investment to end the "chaos" created by local competition for foreign business, the right to negotiate directly with foreigners has been extended to more localities instead of being cut back. The former capitalists involved in the China International Trust and Investment Corporation and various local international trust and investment corporations appear to be playing an increasingly active role in joint ventures, while central and local MOFERT officials sit on the sidelines collecting customs duties.[39]

The most severe intra-governmental conflicts created by the reforms are those between the central agencies and local governments. When the reforms were initiated in 1978, many analysts predicted that the main source of opposition would be middle- and lower-level cadres. Although rural local Party cadres have opposed the de-collectivization of agriculture (see Richard Latham's contribution in this volume), local officials involved in industry and trade have not opposed the reforms that have affected them. In fact, provincial, municipal, and county officials have become enthusiastic supporters of the reform platform, while most of the opposition seems to come from the center. This support of local officials for the reforms stems largely from the decentralization of the fiscal and supply systems, which has enhanced their autonomy vis-à-vis the center. As recent studies by economists such as Barry Naughton and Christine Wong demonstrate, increased control over material supplies, fiscal revenues, depreciation funds, and extra-budgetary funds has given local authorities the resources to act independently of the central planners. And the regulations permitting local enterprises and governments to retain a share of profits, tax revenues, and foreign exchange, have given them the incentive to become economic entrepreneurs. As they became entrepreneurs, seeking out new opportunities for profit and stretching the rules and regulations to retain as much of this profit as they could, local officials became members of the reform coalition. They urge a continuation of decentralization and defend their pet construction projects by arguing that they will "enliven the economy" (a favorite slogan of reform).

When analyzing earlier Chinese reforms of economic structure, Franz Schurmann asserted that administrative decentralization to provincial and municipal authorities ("decentralization II") was in contradiction to economic decentralization to the enterprise level ("decentralization I").[40] In the environment created by the 1978 reforms, which involved administrative as well as economic decentralization, this contradiction may no longer hold. Although administrative decentralization has undoubtedly caused local officials to play a more active role in overseeing the factories in their jurisdiction, the economic incentives offered to local officials under the current reforms make them more solicitous of enterprise managers and more willing to defer to their judgments. The local authorities have become, in a sense, the executives of a local holding company whose interests are closely allied with those of their factory managers. (If economic efficiency has not improved since the arrival of the reforms, it is because neither managers nor local officials have been given the incentive to control costs, not because of a contradiction between administrative and economic decentralization.)

On the other hand, it is obvious that the administrative decentralization component of the economic reforms is a threat to the central Chinese bureaucracy in Beijing. There has been a major campaign to publicize the problems of localism, or what the Chinese are now calling "cooking in separate kitchens." The explosion of local processing factories had created shortages of supplies and robbed the established high-quality consumer goods factories in Shanghai and Tianjin of their usual sources of supply. The local expansion drive has created such a demand for machinery and construction materials that it has interfered with the center's efforts to readjust investment priorities from heavy to light industry. The new local factories are small, inefficient, and wasteful of energy. Local protectionism has prevented the emergence of a national market. Local officials have become so profit-hungry that they engage in fraudulent accounting and reporting practices to cheat the central treasury out of its share. Their greed for foreign exchange has led local officials to smuggle commodities out of China. (In the cases of tungsten and cashmere, two commodities in which China provides a large share of the world market, these uncontrolled exports harmed China's national interests by causing a drop in the international price.) Provinces and cities eager for foreign investment compete with one another to offer the best terms to foreign business people; this competition works to the benefit of foreign businesses and is detrimental to China's interests.

The most serious criticism made by the Beijing authorities against the excessive localism caused by decentralization is that it has weakened the ability of the central state to invest in the energy and transportation infrastructure required for national economic modernization. Even economists who identify themselves as reformers make the argument that recentralization of the fiscal system is necessary if true economic reform is to move forward. In an effort to redistribute financial resources from the localities to the central treasury, Beijing has promulgated taxes on local construction and local extra-budgetary funds, and, as part of the new "substitute-tax-for-profit" policy, has ended the policy whereby localities could retain all the revenues from their locally "owned" enterprises. Even these changes, however, do not totally eliminate the economic incentives for localities to seek industrial profits, and it remains to be seen whether they are sufficient to end localism.

CONCLUSION

I would argue that the move to fiscal re-centralization is not a matter of economic rationality, but is a reflection of the political power of the central bureaucratic state and heavy industry in China. The dispersal of industrial profits, fiscal revenues, and foreign exchange are a "problem" which has several possible solutions. The leaders could institute a "federal" tax system, permitting provinces to set their own tax rates in addition to the national industrial-commercial or income tax. They could look to the provinces to build their own energy and transportation projects; several provinces have already shown themselves willing to finance projects (an airport in Fuzhou, a railroad in Guangdong, roads in many provinces) which will benefit their local development efforts.[41] The central planners could also adjust prices or even eliminate entirely the system of administered prices.

In other words, the answer to the problems created by partial reform could be either further decentralization or re-centralization; the choice is determined by politics and not economics. The re-centralization response to the problems of localism created by the fiscal reform suggests that, despite the addition of local officials—even Communist Party officials—to the reform coalition, the reformers are still unable to prevail.

Despite its size, the reform coalition has not been able to prevent certain critical re-centralization measures or to expand the scope of economic reform. At many points of decision, the reformers have found themselves outmaneuvered by members of the Communist coalition—heavy industry, the inland provinces, and the central bureaucracy. Without fundamental reform of the political system providing democratic institutional avenues for the expression of group interests, the heavy industrial ministries and the other organs of the central bureaucracy will always have the predominant influence over economic policy in a socialist state.

The weakness of the reform coalition also stems from the nature of politics during this era of partial reform in China. Competition among enterprises and localities has been intensified by the new opportunities for economic gain, making the formation of political alliances among them even more difficult than before. Officials concentrate on finding a higher-level patron to protect the interests of their unit and on bargaining with the upper-level organizations rather than on building a united front to influence policy. Until there is a serious effort made in China to restructure political institutions, the power of the Communist coalition and the inability of the reform coalition to act in concert are likely to produce a cycle of reform similar to what has previously occurred in other socialist states.

CHAPTER NINE

False Starts and Second Wind: Financial Reforms in China's Industrial System

BARRY NAUGHTON

When economic reform moved onto the public agenda in China during 1979, there was widespread recognition among Chinese economists that previous experimentation with the economic management system had been unsuccessful. Although some economists simply dismissed previous system changes on the grounds that the economy had never fundamentally departed from a Soviet-type system, more serious discussions stressed that past changes had been dominated by "administrative" decentralizations. Several times in the past, the power of central planners had been drastically reduced, but each time the aim of decentralization had been to increase the management authority of local governments and quasi-governmental management bodies. Since past decentralizations had been carried out without attention to the types of incentives facing local levels, they had led to a lack of coordination in production, excessive and redundant investment, and decreases in the efficiency of both production and investment. The resulting chaos had led in each case to a re-centralization of economic decision-making.[1] Anxious to avoid past mistakes, the economists most committed to reform stressed that future changes in the economic system must be based on a decentralization according to "economic" principles, and that authority relinquished by the central government should be granted to enterprises, and not to local government bodies.[2]

Measured against this standard, the Chinese reform of the past five years has been a failure. Although control over finances, material allocations, and other forms of economic decision-making has been decentralized to a degree unprecedented in the past, only a small proportion of this decentralization has moved according to "economic" principles; the bulk has gone to increase the resources and freedom of action of local government organizations. As a result, serious problems that already characterized the Chinese economic system have been perpetuated and in some cases exacerbated. Chief among these have been the phenomenon of excessive and redundant investment, and the creation of vested interests which lobby against further changes in the economic system. Equally important, the surrender by the central government of a significant portion of its revenues has left government finances in a precarious state, forcing the government into the position of reacting to a series of shocks and crises. Unable to dominate events, the government has had to scramble repeatedly to "put out fires" and prevent disastrous outcomes. The resultant instability in central government policy has significantly impaired the ability of the central government to formulate a consistent program of economic reform.

In spite of these problems, the Chinese political leadership has remained committed to the process of reform. After each shock to the system, the leadership has reaffirmed its commitment to the idea of reform and taken steps to push forward implementation of reform proposals. The central leadership has acted as if they recognize that the problems that have arisen are not the result of "reform" as such, but rather of the failure to implement reform-linked programs in a rigorous and systematic manner. Moreover, in spite of the serious problems that have emerged, there have been substantial accomplishments in the past five years. A program of material incentives that rewards efficient production has been comprehensively implemented by linking bonuses and enterprise benefits to profitability; steps have been taken toward more appropriate costing of factors of production, particularly fixed capital; and the production of consumers' goods is more flexible and responsive to consumer demands. The combination of dramatic decentralization on administrative lines and partial improvements in the types of incentives and information available to local economic agents has dramatically changed the environment in which future reforms must take place.

Initially, central leaders envisioned a series of reforms that would gradually move a relatively centralized economic system toward a partially marketized "mixed" economy. Now, central leaders face a decentralized system in which they must manipulate economic rules and incentives in such a way as to "shove" local agents toward more rational decision-making through a combination of incentives and penalties. In general, this makes further reforms more difficult: Having already given away the "carrot" of greater local control of resources, planners must now wield the "stick" of greater economic accountability. In the meantime, numerous local interests have become accustomed to disposing of a large volume of resources without bearing significant responsibility for the effective use of these resources. Ultimately, the central government will inevitably reassert control over much of the economic realm, but whether it does so in a manner compatible with further economic reforms must remain, at this point, an open question.

In this chapter, I focus on the financial relationship between state-run industrial enterprises and the government bodies to which they are accountable, with a brief look at some of the provisions that allow enterprises to market a portion of their output independently. The emphasis on financial relations is justifiable on a number of grounds. First, from the beginning of the reform process, attention has focused on the development of an appropriate set of rules for dividing enterprise revenues between the enterprise and the state. The Chinese have taken this as the "key link" in the reform process.[3] Second, it has been widely recognized that other portions of the initial reform program—in particular greater enterprise decision-making power in areas such as hiring and firing, democratic management, and a reduction in the scope of materials allocation—have not been implemented to anywhere near the degree to which profit-sharing programs have been implemented.[4] Finally, the changes in the financial system have led directly to changes in the control and direction of investment in the economy as a whole. Most of the funds retained at local levels are ultimately channeled into fixed investment, and, as a result, an understanding of the changing financial relationships in the economy will enable us to understand the collapse of central government control over many parts of the investment process.

ENTERPRISE REFORMS: PROFIT RETENTION

In the five years since 1979, Beijing has promoted three successive "reforms" of the industrial financial system. In each period of reform, a different principle of dividing profits between the enterprise and the government has been advocated, while the proportion of profits actually retained by enterprises has increased steadily throughout the three periods. During the first period, attention focused on the enterprises selected to be "expanded-autonomy" enterprises. Building on experiments begun in Sichuan during 1978, nationwide experimentation was begun with the promulgation of five reform documents on 13 July 1979. These called for selection of a limited number of enterprises to begin:

(1) retaining a proportion of profits
(2) enjoying expanded authority over labor and current production decisions, and the right to market independently output produced outside the state plan
(3) paying fees for fixed capital used
(4) paying interest charges to the bank for all circulating capital used
(5) retaining that proportion of depreciation that had been

remitted to the central government budget, and gradually increasing depreciation rates as production grew.

Of these provisions, the one permitting profit retention attracted the most attention. A profit-sharing ratio was calculated for each enterprise on the basis of the previous year's profits and retained funds. The numerator of this ratio consisted of the bonuses, collective welfare funds, and allocations for new products and expanded production the enterprise had received in the previous year, while the denominator was the previous year's actual profits. Thus, the program was designed in such a way that actual enterprise retentions would increase only to the extent that realized profit increased. Also, enterprises would, in theory, retain fewer funds than previously if their overall level of profits declined; enterprises were required to bear greater risk. Moreover, since the initial enterprises selected for experimentation tended to have very large profits, the retention ratios calculated tended to be quite low, substantially below 10 percent.[5] The initial steps in the reform process, then, were financially conservative.

The conservatism of the first steps of reform conforms closely to the general trend of economic policy in the years 1978-1979, for these years were, on balance, dominated by a *re-centralization* of financial and other kinds of authority. In the aftermath of the Cultural Revolution, central planners moved to regularize and re-centralize control of the industrial management system. This re-centralization included a partial return of enterprises to central control, and a tightening of central government influence over the materials allocation process.[6] On the financial side, the most important measure was a partial re-centralization of depreciation funds. Depreciation is that proportion of enterprise costs which accounts for the "using up" of fixed assets in the production process. In the traditional Soviet model, which China followed in the 1950s, these funds are remitted in their entirety to the central government budget, where they become a major component of investment funds. In China in 1967, however, enterprise funds for local industry were redefined as enterprise resources and ceased to be part of government revenues; in 1971, this provision was extended to all enterprises.[7] These funds, in combination with "flexible resources" provided by statute in local government budgets,[8] provided local governments and local industrial bureaux with a substantial volume of financial resources. Depreciation funds provided the financial counterpart to the dispersed control over materials, which Christine Wong describes elsewhere in this volume for the Cultural Revolution period. In 1978, the central government began drawing 50 percent of all depreciation funds into the central budget, an increase in central government revenues which funded a significant part of the surge in central investment in 1978.[9]

The impact of these changes can be seen most clearly by examining the division of control over fixed investment between the central government and local levels. In Table 21, which consolidates figures for capital construction and replacement and reconstruction investment, the proportion of fixed investment accounted for by local levels is shown declining from 40 percent in 1977 to 34–35 percent in 1978 and 1979. Nor was the central government satisfied with this degree of central control: The government intended to return 10,000 more enterprises to direct central control at the beginning of 1980, although, because of the objections of local leaders, these plans were not realized.[10] This re-centralization in progress was a major source of the chorus of voices raised in 1979 against an overly centralized economic system. Although these protests linked over-centralization with the failings of the Cultural Revolution period, they were in fact directed against contemporary policy decisions (a familiar pattern in Chinese rhetorical practice). At the same time, many enterprises complained because the reform proposals left them with little or no (immediate) increase in financial resources, and these enterprises had to be convinced that the reform provisions would be in their long-term interests.[11] In the meantime, it was repeatedly stressed that profit retention was an experimental program, to be implemented in a small number of enterprises on a provisional basis, in order to gain experience pending full-scale reforms.[12] In order to chart a future comprehensive reform, a working group was convened in mid-1979 under the direction of then Finance Minister Zhang Jingfu.[13] Thus, the initial version of the

TABLE 21 Central and Local Fixed Investment, 1977–1982
(billion yuan)

	1977	1978	1979	1980	1981	1982
Capital Construction	38.2	50.1	52.3	55.9	44.1	55.6
central	(31.2)	(41.7)	(41.3)	(34.9)	(25.2)	(27.7)
local	(7.0)	(8.4)	(11.0)	(20.9)	(19.2)	(27.9)
Replacement and Reconstruction (RR)	16.6	16.8	17.6	18.7	22.5	29.0
central	(1.8)	(2.6)	(4.4)	(3.3)	(3.5)	(3.3)
local	(14.7)	(14.1)	(13.2)	(15.4)	(18.9)	(25.7)
Total Fixed Investment	54.8	66.9	69.9	74.6	66.8	84.6
central	(33.0)	(44.4)	(45.7)	(38.2)	(28.7)	(31.0)
local	(21.8)	(22.5)	(24.2)	(36.4)	(38.1)	(53.6)
Percentage of Total Fixed Investment Under Local Control	40	34	35	49	58	63

Source: *1981 Statistical Yearbook*, p. 295; *1983 Statistical Yearbook*, p. 360; *1983 Tongji zhaiyao*, pp. 58–64; *1981 Economic Yearbook*, IV, 9; *1982 Economic Yearbook*, V, 297; and *1983 Economic Yearbook*, III, 82. See Naughton, "The Decline of Central Control over Industry in Post-Mao China," in *Policy Implementation in Post-Mao China*, ed. David M. Lampton (University of California Press, forthcoming) for a discussion of the categories of central and local investment. Capital construction figures are given according to the augmented definition used from 1983.

Note: Due to rounding, totals will not always be the sum of components.

profit-retention scheme was careful, tentative, and financially conservative.

However, given the extraordinary degree of organizational inertia in the Chinese economic management apparatus, a careful and tentative reform was in danger of being stillborn. During the first nine months of 1979, although the number of profit-retention enterprises expanded to 1,366,[14] these enterprises remained severely constricted in their actual operations. The material supply organs were unable or unwilling to provide materials the enterprise needed, and often unwilling to allow enterprises to market outside-plan output.[15] The organs of the Ministry of Finance continued to watch enterprise finances so closely that their ability to use funds nominally under their control was placed in question. In response to these problems, *People's Daily* on 10 October 1979 published an article entitled, "Be Bold in Taking the First Step in Reform." Contrasting the positive experience of enterprise reforms in Sichuan with the negligible progress made by 8 experimental enterprises in Bejing, Shanghai, and Tianjin, the *People's Daily* commentator stressed that reform was a key policy of the Party center, and that worries about the financial impact of reforms should on no account be allowed to interfere with the implementation of enterprise expanded autonomy. This forceful statement initiated the first period of large-scale change in financial relationships, a period that lasted until December 1980. For the first time, industrial systems brushed aside objections from the Ministry of Finance and implemented enterprise profit retention on a large scale. Change was rapid, and discussion of the ultimate aims of reform reached a new level of sophistication and interest; but many of the economic consequences of this period of rapid change were negative.

In order to overcome the built-in inertia of the management apparatus, reforms had to build up an inertia of their own. The re-centralization of the largest enterprises remained unimplemented, and the reform planning group under Zhang Jingfu faded away, its report never delivered. Most important, the profit-retention program, hitherto experimental and limited in scope, rapidly spread to encompass virtually all the large, profitable industrial enterprises. Detailed uniform regulations for profit sharing were promulgated at the beginning of 1980, and at this time the Ministry of Finance was trying to limit the number of participating enterprises to the approximately 1,600 which had already been approved.[16] In actuality, some 4,000 enterprises were ultimately allowed to keep a share of 1979 profits, and by June 1980 the total participating had reached 6,600, where it was finally stabilized.[17] The profit-retention system changed virtually overnight from a limited experimental program to the predominant financial system in large and profitable state enterprises: The 6,600 enterprises accounted for 60 percent of the output and 70 percent of the profits of in-budget industrial enterprises.[18]

The rapid implementation of profit retention ran far ahead of the other provisions of "expanded autonomy." The difficulties experienced previously in allowing enterprises greater decision-making power over labor and materials continued, while provisions for the compensated use of capital were implemented very slowly. All experimental enterprises were supposed to begin paying for the use of fixed capital in 1980, but in fact only five provinces began implementation in that year.[19] Information is not available about the pace at which enterprises shifted their working capital to bank credits, but it appears that this provision was also implemented slowly. Only the provision that permitted enterprises to retain the full sum of their depreciation retention was implemented with the speed of the profit-retention program.[20] Thus, the rapid growth of profit retention was not accompanied by the adoption of measures that rationalized economic accounting and increased the responsibility of industrial enterprises. And, because of the importance of depreciation retention, much of the decentralization of financial resources was accomplished by simply assigning revenues to enterprises on the basis of the advantages they already possessed (in the form of fixed capital). The "reform" content of profit retention was severely limited.

In spite of the demise of a comprehensive reform program, considerable sentiment existed during 1980 for pushing ahead in the direction established. During the National People's Congress meeting in August-September 1980, it was announced that profit sharing would be expanded to all state enterprises by the end of 1981.[21] But, just as this decision was made, a real financial crisis was developing. From June 1980, profit remittances to the state began to fall below the previous year's figure in each month, and in August the sum was fully 17 percent below the previous year.[22] The financial crisis the Ministry of Finance had always warned of was actually emerging. Under these circumstances, complaints about rising prices which had been voiced during the National People's Congress were paid greater attention. A large-scale budget deficit was shaping up which, unlike the 1979 deficit, could not be blamed on extraordinary circumstances like the war against Vietnam, or the costs of terminating projects begun during the 1978 investment boom. Planners also were alarmed by the explosion in the amount of currency outstanding, which they considered to be a sign of serious inflationary pressures.[23] As a result, in December 1980, Beijing halted the expansion of enterprise reform experiments, enacted strengthened price controls on consumer goods, and drastically reduced spending. Thus ended the first period of enterprise reform. From Table 21, it can be seen that the proportion of investment accounted for by local agents (including enterprises and local governments) took a jump upward in 1980 to 49 percent, surpassing the 1977 high by a substantial proportion. From 1980, it also becomes possible to account roughly for the financial sources of local investment. This is done in Table 22, where it can be seen that, in spite of the rapid spread of profit retention, retained depreciation funds remained the largest funding source by a wide margin. This has remained true since, and represents a major element of continuity between the Cultural Revolution period and the current post-reform situation. Bank loans represented the other rapidly growing component of local investment, perhaps tripling between 1979 and 1980. Although these loans bore interest charges, rates were low, and pay-back provisions quite liberal. As a result, it must be concluded that the expansion of bank credit was also a very low-cost addition to local-level financial resources.[24]

For the first three months of 1981, enterprise reforms were put on hold as the government implemented a drastic deflationary policy. Planned budgetary investment was slashed by 40 percent; some enterprise bank deposits were frozen; and the new program of bank loans for fixed investment was suspended. This was strong medicine, and it quickly became apparent that the deflationary policies were enlarging the government deficit they had been intended to reduce, as production—and still more so profits—took a downward plunge. By the end of March, most of these measures had been scaled back, and, around September, plans were readjusted upward and the expansion of heavy industry resumed.[25] However, the impact of this brief episode was destined to be far-reaching; although at the time numerous observers perceived this period as one of re-centralization,[26] the ultimate impact was just the opposite. As the regime labored throughout 1981 to undo the harmful effects of its

TABLE 22 Total Local Funding Sources, 1980, 1981, 1982
(billions of yuan)

	1980	1981	1982
1. Depreciation	18.3	20.0	22.0
2. Retained profit	6.2	7.7	11.1
3. Domestic loans	8.1	12.3	[15.7]
4. Local government and non-profit organizations	7.8	8.3	8.8
5. Foreign capital	[0.4]	[0.7]	[1.6]
Total	40.8	49.0	59.2
Less bond purchases	–	–4.9	–2.2
Revised total	40.8	44.1	57.0
Realized investment	36.4	38.0	53.3

Source: Naughton, "The Decline of Central Control." The contribution of profit retention to investment finance is calculated on the grounds that 60% of retained profits are channeled into fixed investment.

deflationary policies, it again endorsed programs that would allow enterprises to retain profits. In the spring of 1981, a second period of change in enterprise financial relations began, centering around the system of "profit contracts" (*yingkui baogan*).

PROFIT CONTRACTS

Profit contracting grew directly out of the unusual conditions created by new central government policies in early 1981. Drastic revisions in enterprise plans took place at the beginning of 1981, and central planners in their haste unavoidably submitted a contradictory mixture of targets to the enterprises and local industrial bureaux. Output targets, and the corresponding government purchase agreements, were substantially lowered, following the cutback in government investment. These targets are disaggregated to the enterprises on the basis of functional systems, that is, through the central industrial ministries. Profit targets, on the other hand, although revised, remained high, and these targets were disaggregated to the enterprises on the basis of the government units (provinces or cities) to which the enterprises are financially subordinate. Coordination of these plans is difficult in the best of circumstances, and, not surprisingly, in a period of rapid change there were substantial conflicts. At the end of January, the center was just beginning to send revenue targets to the local levels and, as one might expect, these were meeting much resistance.[27] Enterprises reported that their profit targets were impossible and refused to accept them. Moreover, some factories experienced severe morale problems when they discovered they had literally no work assigned to them. The production plan for enterprises under the First Ministry of Machine Building were set at only one quarter of production capacity.[28]

Profit contracting developed as an ad hoc response to this situation. Under this system, the enterprise and its supervisory body negotiated a "base figure" of profits which the enterprise undertook to deliver to the state. Enterprises were then allowed to retain a very high proportion of profits above this base figure, ranging from 20 to 100 percent, with the retention rates frequently increasing with the degree of overfulfillment.[29] From its initial introduction in Shandong province in the spring of 1981, this system spread rapidly. By August, 65 percent of in-budget state enterprises had adopted the profit contract or other system of profit retention, and, by early 1982, this figure had risen to more than 80 percent.[30] Whereas profit-retention ratios in the initial reform period had been based firmly on the previous year's performance, such was not the case with the profit-contract base figures. These were set on the basis of negotiation, and on "seat-of-the-pants" calculations of what the enterprise might be capable of. The idea was to develop an optimal "striving target" with high retention beyond this target in order to mobilize enterprises to fulfill their responsibilities in a difficult situation. Since the center was the immediate cause of the instability in the economic environment, it was unfair to ask the enterprises to bear the risk; as a result, rewards were increased for those who fulfilled their targets.

The flexibility inherent in the profit-contracting system had other obvious advantages to China's industrial managers. Plan revisions were not the only source of financial pressure on enterprises. Prices of nearly all industrial raw materials had been increasing since 1979, while prices of many kinds of manufactured goods—especially machinery and consumers' durables—had been declining. Moreover, enterprises were being required to pay more for environmental protection and insurance, while increased wages and bonuses were also putting pressure on costs.[31] In some cases, profit contracting simply became a tool for developing a unified financial and investment plan. After calculation of the year's investment plan in a given enterprise, the profit base figure would be adjusted to permit the enterprise to retain at least this sum of money.[32] These advantages were summarized in a *People's Daily* article of August 5th which called for widespread implementation of this system:

> Profit contracts by sector (that is, industrial bureau) can not only solve the difficulties created for the enterprise by the lack of coordination between production plans and profit plans, but also allow adjustment of revenues within the sector, thus solving the inequities created by profitability differentials between enterprises; it is also advantageous to unified long-range planning... and creates the conditions for organizational and structural changes.[33]

It should be obvious, though, that the advantages outlined in the previous comment are the advantages of a planned economy, and not those of a true "reform" program. The profit-contract system was to reveal other defects as well.

The most obvious failing of the profit-contract system was the lack of an objective standard on which to base the amount of profit an enterprise was allowed to keep. In an unstable economic environment, base figures could not be based on the previous year's profits, and an enterprise with lower targets or falling sales or prices might reasonably demand a lower base figure. The need for worker housing, more materials, or any factor, real or invented, could be used to justify the least profit remittance to the state. Profit base figures became the focus of negotiation between the enterprise and its superordinate body. Nor did the negotiability of targets cease once a base figure had been selected. Since the essence of the system was to respond "fairly" to factors outside the enterprise's control, those enterprises which performed poorly in the course of the year were quick to find "objective" factors for their poor performance. In one of the model districts for the profit contract system, Yantai in Shandong, a retrospective investigation revealed that, although 46 profit-making enterprises failed to attain their base figures, and several losing enterprises showed losses greater than their contracted loss subsidy, only 2 enterprises were actually penalized by deductions from enterprise funds.[34] Nor was this an isolated example: The four provinces of Shandong, Liaoning, Beijing, and Tianjin, which implemented profit contracting in a particularly thorough fashion, together fell short of their revenue targets by over 1 billion yuan.[35] Chinese critics of the profit-contract system labeled it a system where enterprises were "responsible for profits, but not for losses." Although the system was supposed to increase the ability of planners to deal with extraordinary circumstances, it had the paradoxical result of increasing the bargaining power of all enterprises. Even those enterprises remaining on the former profit-sharing program could argue that extenuating circumstances required specially favorable treatment.[36]

As a result, total profit retained in enterprises increased substantially in 1981. This increase was particularly marked because, after September, plans were adjusted upward and industrial production grew rapidly. But base figures had been negotiated early in the year on the basis of lower plans, and they were not readjusted. As production expanded in the last quarter of 1981, and continued expanding through 1982, many enterprises were able to retain large sums of money. While failing enterprises escaped responsibility for their losses, enterprises that hooked into the rising output trend were able to reap windfall gains. The development of total profit retentions is shown in Table 23, and the trend of an increasing volume of local financial resources can be clearly traced in Tables 21 and 22 as well. Local investment amounted to fully 58 percent of the total in 1981, and 63 percent in 1982. The degree of decentralization of financial resources had far surpassed the highest past levels.

Yet, this decentralization had been accomplished without a corresponding progress toward a real reform of the industrial system. Indeed, as indicated above, the profit-contract system actually intensified the existing administrative chain of command in the economy. In a sense, the establishment of profit contracts with enterprises was simply an extension down to the

TABLE 23 Profit Retained by State-Run Enterprises, 1978-1982
(billions of yuan)

Year	Amount
1978	2.1
1979	6.7
1980	9.6
1981	11.8
1982	17.0

Total Profit Retained 1979-1981: 28 billion
of which:

Enterprise funds	4.9
Profit sharing	14.2
Profit-loss Contracts	7.0
Tax for profit	2.0

Sources: 1978, 1981: Wang Bingquian, *Jingji guanli* 11:3(1982).
1982: Wang Bingquian, in BBC FE 7209 C2/3.
Breakdown of 1979-1981 Total: Shen Jingnong and Tao Cengji, *1982 Economic Yearbook*, V, 322.
1979-1980: Naughton, "The Decline of Central Control."

enterprise level of the way profit targets were disaggregated in the economy as a whole. The central budget was set up in conjunction with a projected profit total; this total was then disaggregated down the chain of command. Targets were set for the province, city, municipal industrial bureau, and finally the enterprise, with each level being responsible for the targets and target fulfillment of the units immediately subordinate to it.[37] The profit-contract method was extended uniformly from the very top to the very bottom of the system. In such circumstances, local government and industrial management bodies naturally intensified their interference with the decisions and activities of subordinate enterprises. After all, one of their most important responsibilities, that of revenue collection, depended directly upon enterprise decisions. Local governments were encouraged to act as revenue-maximizing corporate bodies, even though irrational prices and other structural rigidities implied that such maximization would come in conflict with overall economic objectives. As a result, while a greater proportion of financial resources was disposed of at the lower levels of this administrative chain, the fundamental determinant of how revenues were divided remained the interactions among the various levels of the hierarchy. Bargaining over profit became one of the main activities of the industrial hierarchy, replacing bargaining over plan targets.

As the defects of the profit-contract system became increasingly clear, a debate began about the value of this system. Although this debate was ultimately to lead to a drastic downgrading of the importance of profit contracting, and the supersession of this system in a third period of financial reforms, for a substantial part of 1982 advocates of the profit-contract system seemed to dominate policy. The course of this debate can yield some additional insights into the industrial management system. Just as the initial profit-retention experiments were concentrated in the largest enterprises, the profit-contract system was implemented primarily among the smaller enterprises. (Although some profit-retention enterprises switched over to the new system, there were still "over 6,000" profit-retention enterprises at the end of 1982.) The smaller enterprises—and a good number of the large enterprises as well—are under the direct control of local governments. As Christine Wong describes elsewhere in this volume, the output of these enterprises is already largely controlled by the local governments. Under the profit-contract system, local governments were essentially free to manage local industrial systems as they saw fit, without interference from the central government (beyond the obligation to fulfill their own profit contracts), and without a strong obligation to institute any of the reformist programs that had been included in the initial reform documents. By combining material and financial authority in the hands of local governments, the system as it evolved in 1981 intensified the tendencies already present to develop local industrial systems with only limited links to the national economy. Tendencies toward local "protectionism" developed in this period and were persistently criticized by central planners.

There had always been a strong current of thought in China, however, that advocated a kind of "dual track" approach to planning problems. In this approach, market-oriented reforms would be instituted for small enterprises (and particularly small consumer-goods producers), while the state would intensify its control over the largest enterprises. The vital task for central planners would then be to improve the quality of planning insofar as it related to the largest enterprises, while local industries would be free to go their own way, subject to an increased degree of market discipline.[38] The initial development of reform had proceeded in the opposite direction: The largest enterprises had rapidly adopted profit-retention programs. But this was not because of the design of the reform program (which we have seen was left behind by the rapid development of events). Rather, it was because the industrial ministries (which might be supposed to be the very people most interested in strengthening central control over large enterprises) perceived that their enterprises would dispose of more resources (through depreciation and profit retention) if they adopted the "reform" package, and so moved rapidly to take advantage of this opportunity. These forces perceived in the profit contract system an additional opportunity to intensify their control over the larger enterprises.

Initial approval of profit-contract systems had been extended only to enterprises with very low initial levels of profits,[39] but the program rapidly spread beyond these limits. Beginning roughly in August 1981, under the leadership of Yuan Baohua of the State Economic Commission, a campaign was begun to promote "responsibility systems" in industry, enterprise rectification in the larger enterprises, and the experience of the Capital Steel Company. An unusually efficient large-scale enterprise, Capital Steel had been named a profit-sharing enterprise in 1979, and, in the second half of 1980, was shifted to the more advanced "tax-for-profit" system (see below). During 1981, it was transferred to an especially rigorous version of the profit-contract system, implemented a thorough program of assigning work quotas to individual workers (and linking their bonuses to their quota fulfillment), and was put forward as a model of the industrial "responsibility system."[40] Thus, while *responsibility system* is a broad term (with a long history in China) which can cover any financial system in which rewards are linked directly to performance,[41] the role of Capital Steel as the prime embodiment of the virtues of such a system ensured its practical identification with the program of profit contracts. Simultaneously, the program of "enterprise rectification" was begun. This program, directed specifically at the largest enterprises, aimed to regularize leadership groups and work assignments, ideally by establishing pervasive work quotas and profit contracts in the style of Capital Steel. The essence of these moves, then, was to strengthen planning, and bring the largest enterprises under the profit-contract system.

These moves were quite controversial, particularly inasmuch as the Capital Steel experience was advanced as a variety of, and continuation of, "reform" of the economic system. Reformist economists alternated between rejecting the Capital Steel model, since it was not really reformist, and claiming that the real reason for its success was the expanded decision-making power it had gained along with its status as an exemplar.[42] Criticisms of profit contracting multiplied throughout 1982.[43] Ultimately, though, it was the weaknesses of the program itself, and the unwieldy nature of the administrative apparatus that spelled the end of the reliance on profit contracting as the central direction of financial-system changes. As it became clear that the total amounts of profit retained under the profit contracting system had significantly exceeded previous totals (see Table 23), the program came under fire for its impact on government revenues. Simultaneously, the reinforcement of local independent systems was increasingly being seen as a critical problem. And, finally, the rectification program itself began to run far behind schedule, as enterprises and their superordinate bodies deflected changes that would tighten up their operating conditions and subject them to a more rigorous system of rewards and penalties.[44] In the end, only a few large enterprises were shifted to the profit-contract system,[45] and, by the end of 1982, the way was prepared for the third phase of change in industrial financial systems.

TAX FOR PROFIT

The focus of the third period of financial reform is the "tax-for-profit" (*li gai shui*) system. In its fully developed form, this system involves substituting a series of taxes, paid directly to state treasuries, for the previous system of profit deliveries channeled through the enterprise's superordinate body. These taxes generally include:

(1) charges on fixed and circulating capital
(2) the existing sales tax ("industrial-commercial tax")
(3) an income tax
(4) an "adjustment" tax to equalize the burden between different enterprises with more or less favorable operating conditions (including prices, location, and natural resource endowment).

This system had been tried experimentally since 1979. Originally tested in Guanghua county, Hubei, for regulating small county-run enterprises, it was expanded to 191 enterprises during 1980, including some large ones (one of which was Capital Steel). In spite of the predominant attention given the profit-contract system during 1981-1982, the tax-for-profit experiment expanded modestly during this period, covering 456 enterprises by mid-1982.[46]

The crucial difference between the tax-for-profit system and the other systems of profit retention is that, in the tax-for-profit system, an attempt is made to charge the enterprise realistically for all factors used in the production process. In theory, this permits the system to function with a degree of automaticity (insofar as this is possible without flexible prices), and designates the enterprise as the residual claimant on net income. Thus, if the various taxes and charges can be set at reasonable levels, the opportunities for bargaining over profit targets and retention rates should be drastically reduced. Questions of fairness become less upsetting to the financial decision-making process, and enterprises can be forced to be more accountable for their losses as well as profits. Numerous problems stand in the way of an effective implementation of this program (not the least of which is a rigid and irrational price system), but at least the direction of change is one that can be built upon in further reforms. It is necessary to examine the extent to which actual implementation has realized the promise inherent in this system.

We noted above that implementation of fixed capital charges had proceeded very slowly through 1980. In the years since, further implementation has continued to be slow and uneven, but it appears that progress has nevertheless been made. Central directives in early 1981 decreed that all "expanded-autonomy" enterprises begin paying capital charges, and, in early 1982, it was decreed that all industrial enterprises should begin paying such charges.[47] But these directives are still rather far from being fully implemented, although precise figures are not available.[48] Implementation has been slowed by several factors. Many enterprises don't make enough profit to be able to afford fees at reasonable levels; many enterprises have just been completing capital inventories through 1980 and 1981; and enterprises have resisted (sometimes successfully) the imposition of a new financial burden. But, in some areas, the program has been widely applied. Shanghai is the leader.

Shanghai charges the highest rates for fixed-capital occupation, 9.6 percent annually. (Sichuan is the lowest at 2.4 percent.) Moreover, in Shanghai (as in Tianjin and Beijing, but not other provinces), capital charges are paid by the enterprise out of its own funds, *after* its profit share has been calculated (or after the exaction of income tax, depending on the system). Thus, enterprises in Shanghai have very real incentives to economize on their use of fixed capital.[49] Finally, coverage in Shanghai has been broad. As early as 1980, Shanghai enterprises paid 496 million yuan in fixed capital charges, equivalent to 5.8 percent of the total fixed-capital stock (underpreciated value). This would imply a coverage of 60 percent of the capital stock, assuming that the charges were applied to underpreciated value of capital in that year (or nearly total coverage if, as was the case beginning in 1981, the charges were applied to depreciated value).[50] A major change in the incentive environment facing Shanghai enterprises has been successfully implemented.

There are some additional interesting aspects to the capital-charge system formalized in 1981. After some debate on whether charges should be exacted on depreciated or underpreciated value of capital, it was decided to charge enterprises for the depreciated value of their fixed capital as of the end of 1980, and for the undepreciated value of new assets financed by government allocations or bank loans after this date. When enterprises remit depreciation to the central government, this will be counted as a reduction of their capital stock, but the depreciation funds retained by the enterprise will not be so counted. Investments made by the enterprise itself, out of retained profits or depreciation, will not be subject to the capital charge; they will be, in a limited sense, the "property" of the enterprise. The enterprise will, however, be responsible for paying capital charges on the full value of government-financed investment indefinitely (in the absence of depreciation remittance), regardless of whether it has replaced government investments with its own investments or not, because, along with depreciation retention, the enterprise has accepted the responsibility of maintaining the productive potential of government-financed assets at at least their original level. Thus, the effective control of the enterprise over its own capital stock has been greatly enhanced, while its obligations to the government remain in force.[51] This is an ingenious and workable system.

The implementation of capital charges is a precondition for meaningful implementation of the tax-for-profit system. In this sense, the gradual spread of capital charges created the conditions that allowed the government to shift to advocacy of the tax-for-profit system at the end of 1982. At the same time, the slow and uneven spread of capital charges has continued to impede the full implementation of this system. At the beginning of 1983, it was decreed that *all* enterprises would shift to the tax-for-profit system beginning on 1 July 1983. Inevitably, however, only a somewhat stunted version of the system could be implemented on such a broad scale. Pared to its essentials, the 1983 tax-for-profit system simply requires enterprises to pay a flat-rate 55 percent income tax (or a graduated income tax in the case of enterprises making annual profits of less than 200,000 yuan). Profit-retention ratios or profit-contract base figures are then to be recalculated on the basis of after-tax income in such a way as to leave most enterprises with the same level of retained profits.[52] Thus, the shift to tax-for-profit in 1983 will not drastically alter the enterprise financial system at a single stroke. But its importance lies in the fact that it establishes a future direction of change that is compatible with further reforms: As capital charges are made universal and prices rationalized, the scope for bargaining will be reduced and the tax-for-profit provisions can become increasingly operational.

Actually, the immediate advantages of the tax-for-profit system lie in its impact on the role of local governments. First, the new taxes are to be shared between the central and local governments, and they thus represent a net increase in central government revenues, and a net decrease in local government revenues. This is rational and in line with central government attempts to reduce deficits at the central level. Second, the new system marks a first step in the attempt to reduce the tutelage of the enterprise to local government organs. Since the reduced amount of profit can be increasingly earmarked as the specific revenue of the enterprise, it marks an improvement over the previous system where the larger amount of profit was basically under the control of the local government, which then undertook to share it with the enterprise. This may have an impact on the amount of day-to-day interference in enterprise management by local government authorities (although fundamentally this change can come about only as the result of more thoroughgoing changes in the industrial system). Thus, the tax-for-profit system is an attempt to reduce the extent to which enterprises are subject to a kind of local government ownership, and create the conditions for greater enterprise autonomy. Whether the tax-for-profit system can realize its potential in the future reform of the Chinese industrial system will of course depend crucially on the future political evolution of China, and the continued commitment of the regime to thoroughgoing reforms.

APPRAISAL

In a very real sense, reform of the financial system in Chinese industry has just begun. And yet, the net result of five years of continued experimentation with enterprise finances has been a progressive, uninterrupted increase in the amount of profit retained by enterprises, and a corresponding decrease in the amount of profits included in central government revenues. This can be clearly seen in Table 23. During the period when profit contracting was being stressed, and in which it could be argued that true reforms were marking time, this decentralization continued unabated. Yet, even more telling is the fact that retained profit has remained a fairly small proportion of the total amount of financial resources available at local levels. As inspection of

Table 22 reveals, the largest component of local financial resources has continued to be retained depreciation funds. Nor has this situation changed in 1983. The government did exact a 10 percent levy on all "extra-budgetary funds," which includes retained profit and depreciation as well as the subsidiary resources of local governments and non-profit organizations.[53] But, since these funds have been growing substantially more rapidly than 10 percent per year, it is unlikely that this modest level will fundamentally alter the situation. Barring some drastic change of direction in economic policy-making, China will have to cope with a highly dispersed pattern of investment financing for years to come.

Even within the quantity of retained profits, most such profits are left with the enterprises either as the outcome of a bargaining process with their superiors, or automatically on the basis of the total wage bill (as in the enterprise-fund system). In 1980, the one year for which we have a comprehensive breakdown, enterprises participating in the profit-retention program (which could at least lay partial claim to being a reform program) retained 1.24 billion yuan more than they would have retained in traditional, unreformed provisions.[54] This is a small proportion of total profit retentions in that year, and a very small proportion of total decentralized financial resources. The conclusion is unmistakable, then, that the dramatic decentralization of finances that has occurred in China since 1979 is primarily an "administrative" decentralization. Revenues have been assigned to local governments and industrial bureaux on the basis of their capital stocks (depreciation), or their bargaining power, but rarely on the basis of their economic performance. This decentralization, then, should be seen as the latest in a series of such decentralizations to have occurred in China since 1958. Not surprisingly, serious problems relating to the coordination of fixed investment have followed this decentralization, just as such problems followed all previous decentralizations.

The fundamental problem is that too much money has been invested in too many different investment projects. As a result, an enormous quantity of resources has been tied up in uncompleted projects. This has been a problem since at least the early

1970s. During 1973-1975, strenuous efforts were made to reduce the number of projects under construction, but these efforts were not successful.[55] This unsolved problem was greatly exacerbated by the large-scale initiation of new projects in 1978, many of which could not be readily canceled because they involved contracts with foreign countries. By the end of 1980, the staggering sum of 137 billion yuan was tied up in the investment process. Of this total, 80 billion consisted of uncompleted projects, and 22.5 billion of stockpiles of machinery and equipment.[56] This total amounts to almost double the total fixed investment in that year, and represents an enormous quantity of physical resources which are making no contribution to increased production.

Since 1979, Chinese planners have attempted to address this problem. On the one hand, they have concentrated on reducing the total number of projects under construction by canceling or delaying projects that are impractical or currently unneeded. On the other hand, they have attempted to reduce the flow of annual investment, in order both to release resources for consumption and to concentrate the flow of investment on more effectively planned projects. These initiatives have not been successful; in part, they conflict with each other, and, furthermore, serious obstacles have blocked their effective implementation.

A large number of projects have simply been suspended. Of 1,624 active large-scale investment projects at the end of 1978, almost 900 had been abandoned or suspended by the end of 1981. The total outstanding as of that date was 663, with only 289 having been completed in the previous three years, and 200 new projects begun.[57] While the suspension of projects is useful, since it leads to the concentration of investment funds on a smaller number of projects, it does not provide a way to bring resources already sunk in suspended projects into play. Ultimately, only new investment that complements resources already committed can draw those resources into the production process. In the case of large-scale investment projects, this new investment must inevitably come from the central government. It is here that the conflict with the second objective—reducing overall investment—becomes acute.

From Table 21 it can be seen that, while total fixed investment has been increasing since 1978 (with the exception of 1981), central government investment has declined substantially. The decentralization of financial resources since 1979 has, however, led to a dramatic increase in the total quantity of locally funded investment. This has prevented the effective implementation of cutbacks in total investment initiated by the center, and greatly intensified the dilemmas faced by central planners. The basic situation can be simply described: In a situation where both consumption and local investment are increasing rapidly, the central government has been forced to bear the entire burden of investment cutbacks, in order to maintain a modicum of macroeconomic balance. With the flow of central government investment drastically reduced, the ability of the government to bring resources already committed in existing investment projects into the production process is seriously limited. Little progress can be made in reducing the real stock of unusable resources. Furthermore, the ability of the central government to direct resources into key bottleneck sectors—particularly energy and transport—has also been crippled. In spite of widespread recognition that energy production is the key bottleneck in the Chinese economy, investment in energy actually declined after 1978, before turning up slightly in 1982.[58] Investment in energy and transport infrastructure is necessarily large-scale, which only the central government can effectively provide. Moreover, since central government investment is used to "balance" the macroeconomy, it is subject to drastic changes, as in the opening months of 1981. This instability in central government investment policy creates instability in the whole planning network, and disrupts the stable environment necessary both for effective planning and for the further progress of reforms.

During the first "high tide" of reform, during 1980, it was widely held by reformist economists that the fundamental cause of the over-extension of investment was the practice of providing investment funds without charge. Reformers believed that by charging for capital the "over-extended investment front" would rapidly be brought under control.[59] As we have seen in the preceding pages, this proposition has not yet been adequately tested in China, since most investment funds continue to be made available without adequate accountability. Experience in other socialist countries, however, particularly Hungary, seems to indicate that charging for capital is not in itself sufficient to check the forces that create excessive investment demand. In particular, Janos Kornai has argued that socialist enterprises, even when they pay for capital, are not subjected to serious threats of bankruptcy or punishment if investments turn out poorly. Due to the absence of effective sanctions for inefficient investment (what Kornai calls a "soft budget constraint") and to the involvement of enterprise managers' careers in the effective expansion of enterprise operations, enterprises continue to demand high levels of investment, a phenomenon Kornai labels "expansion drive."[60] Currently, it is absolutely unquestionable that Chinese enterprises face a soft budget constraint: Numerous avenues exist for enterprises to escape the consequences of misguided decisions in investment or production. The operation of the profit-contract system practically exemplifies the meaning of a soft budget constraint, and Chinese economists describe the same phenomenon when they say that enterprises are "responsible for profits, but not for losses." While the tax-for-profit system may effect some marginal changes in this situation, it is unlikely to alter things fundamentally in the foreseeable future. This means that central planners will have to continue to resort to administrative controls over the types of investment projects allowed; indeed, in 1983, major steps were taken to strengthen these administrative controls by enacting intensified supervision of particular kinds of machinery and raw materials, and by setting up provincial investment commissions to eliminate as many nonessential investment projects as possible.[61] The continued need for administrative controls stands as a major obstacle to continued progress in reforms.

While the dispersion of investment created by the most recent wave of decentralization measures is similar to the problems created by past decentralizations, there are also some new elements. Chief among these is the change in the types of investments local agents are engaging in. There has been a massive shift in investment away from goods-producing capital and toward the construction of housing. The major funding source for this housing boom has been the retained profits of enterprises.[62] While the social effects of the increased supply of housing have been highly beneficial, this change in the composition of investment has put tremendous strains on the supply of building materials. The proportion of capital construction going for construction and installation, which had remained below 60 percent until 1978, reached 72 percent in 1981 (with the proportion spent on machinery declining from over one-third to only 19 percent). As a result, the cost of new construction has been increasing at a rate of 9 percent per year since 1978, the highest rate of inflation in any sector of the Chinese economy.[63] Moreover, state key-point investment projects have experienced difficulties acquiring adequate supplies of building materials, and these shortages have been blamed for the failure to complete a number of these projects on schedule.

In past decentralizations, local investment has been concentrated on the development of heavy industrial products needed to assure local self-sufficiency, such as coal, iron, cement, and simple machine tools. In the most recent phase, this investment has been concentrated on consumer goods which yield high rates of return in terms of profit and tax. One particularly egregious example in this area lies in the production of electric fans, which, due to the quirks of the Chinese pricing system, happen to be an extremely profitable item. In Wuhan before 1978, there was a single factory producing electric fans. By the end of 1979, there were 23 factories producing fans in Wuhan, and over 1,500 in the country as a whole.[64] Since the same irrational high price which makes fans highly profitable also restrains demand, fans rapidly became an item in excess supply. Similiar examples could be cited for most consumer durables, as well as synthetic fabrics, cigarettes, liquor, and other lucrative items.

Not all the factories newly producing consumer durables are the creation of local investment projects. Many of these new producers were formerly producing machinery for government investment projects. After 1979, investment demand for their output declined sharply. Simultaneously, however, the reform provisions of 1979 allowed these factories to market output which they produced outside the plan. While these provisions were widely credited in the Chinese press with allowing machinery enterprises to remain in operation, and even occasionally turn a profit, it should be noted that their ability to continue producing is directly related to the collapse of central government control over investment. On the one hand, outside-of-plan sales have been a major source of machinery for local investment projects; on the other hand, factories have shifted production to consumer durables. Thus, oxygen-machinery plants wind up producing fans, foundries produce furniture, and so on.[65] While the opportunity cost of using idle machinery factories to produce consumer durables may appear to be low, given the limitations on demand for high-price items, these provisions only postpone the day of reckoning. As these goods flood the market, machinery enterprises must find a new direction for development. In the meantime, enterprises producing increasingly demanded construction materials—particularly cement and lumber—found the provisions permitting direct factory sales to be a significant temptation. In 1981, the 53 large- and medium-scale factories producing cement sold, on their own, 2.6 million metric tons of cement, and in the process underfulfilled their state sales plan by almost that identical amount, resulting in a 10-percent shortfall.[66] Similar situations were reported with regard to lumber, steel, sewing machines, and woolen fabrics. In order to protect its own investment projects, the central government began, in 1981, to restrict the right of certain enterprises to sell outside of plan output.[67] In 1983, these provisions were extended into a general regulation that only a portion of outside-of-plan production could be sold by the enterprise itself, and a portion was to be sold directly to the state materials allocation agencies.[68]

As of mid-1983, renewed administrative controls over investment had completely failed to halt the expansion of the number of local-level investment projects. There were some 45,000 small-scale investment projects underway at the end of 1981; during 1982, 33,500 new projects were initiated, and 70,000 projects were still active at year-end. In the first six months of 1983, an additional 10,716 new projects were begun.[69] At the end of June 1983, the Chinese government, in conjunction with the Central Committee of the Communist Party, sent down an urgent directive on controlling the scale of investment. Each province was directed to reduce its total investment spending, and steps were taken to ensure the flow of raw materials to state key-point investment projects. Undoubtedly these measures will have some short-run efficacy, but where they will lead in the future is difficult to predict. The problem of investment control will be persistent and difficult for Chinese planners in the years to come.

CONCLUSION

Given the serious problems with investment control in contemporary China, why do central planners not simply re-centralize a substantial portion of financial and material resources? The answer is twofold. First, planners recognize that drastically to re-centralize financial resources would be to destroy the foundation that has been laid thus far for a comprehensive economic reform. A whole new system has been set up on the basis of the new financial provisions, and there is no simple substitute for the existing system. Second, planners recognize that they simply do not have the technical and informational resources to effectively plan the entire economy. Even if control over all resources were concentrated in the State Planning Commission, it would be unable to deploy these resources rapidly and effectively. In this sense, the costs of investment dispersal may not be as severe as they appear. So long as central planners can assure the supply of materials for the limited number of projects they *can* plan effectively, the current system may not be the worst of all possible systems. The central government remains committed to a long-range reform of the management system.

Yet, it must be recognized that recent developments severely jeopardize that reform, in spite of the progress that has been made. Local governments now enjoy unprecedented control over resources, and they actively resist attempts by the central government to reduce that control, or make local governments more accountable for their decisions. As a result, local power-holders who enthusiastically supported reforms when they increased their resources may now oppose programs like the tax-for-profit system and the institution of capital charges, which limit their maneuverability. The central government may be able to compel local government acceptance of the changes they envision, particularly given the ongoing top-down purge of Communist Party membership. But the process will be more difficult as consensus becomes harder to achieve.

At the same time, it is imperative that central planners exert some control over the investment process. Since they are essentially limited to administrative means for doing this, they will inevitably take actions that contradict the ability of enterprises to determine their own investment decisions, based on their own financial resources. This will remain a fundamental contradiction with the progress of reform. In a sense, then, real financial reforms are just beginning. But whether they will survive and take hold remains a difficult question. The first real indications should come as the government begins to implement a comprehensive price reform, which could begin in early 1985. The ability of the government to push an unwieldy, bureaucratic system toward greater economic rationality will then be put to the test.

CHAPTER TEN

Material Allocation and Decentralization: Impact of the Local Sector on Industrial Reform

CHRISTINE WONG

A key objective of industrial reforms in the post-Mao period is to increase efficiency in resource use. Among the primary causes of the deterioration in capital efficiency indicators during the Cultural Revolution were over-investment and severe sectoral imbalances. Bottlenecks in transport and energy supplies, along with chronic and pervasive shortages in a variety of producer and investment goods, had caused long delays in construction projects and substantial underutilization of existing plant and equipment.[1] Believing that these problems were products of the command economy and the habit of "eating out of the communal rice pot," Chinese planners have moved to inject greater flexibility into the system, alter the incentive structure guiding interaction among economic agents, and give a greater role to market mechanisms in coordinating resource allocation. To this end, reform measures reduced the scope of central planning and material allocation, cut the number of commands sent to enterprises, gave them greater autonomy in making production and investment decisions, and simplified wage and bonus regulations. To make enterprises more cost-conscious in their operations, a variety of profit-retention schemes have been introduced, interest charges on capital are now levied, and bank loans have increasingly replaced budgetary allocations in financing investment. All these serve to decentralize decision-making authority and to induce local governments and production units to take greater responsibility for their actions.

Many of these reform measures are familiar to students of Soviet-type economies. The problem faced by planners in China and elsewhere was how to decentralize and give greater autonomy to microeconomic units while retaining essential control and ensuring that social or ideological constraints are not violated. This chapter looks at the process of decentralization in China from the perspective of control over material allocations. Of particular interest are the extent to which the Chinese economy has been decentralized in the post-Mao period, the nature of this decentralization, and its implications for economic performance.

Allocation of key producer and investment goods is a crucial mechanism of control in a planned economy. In a classic command economy, where planning relies on material balancing, material resources have a significance far exceeding that of financial resources. According to China's foremost economist, Chen Yun, "How much capital construction we can undertake is not determined by how much money we have but by how much we have in materials."[2] During the post-Great Leap Forward economic crisis in the early 1960s, when the government undertook drastic retrenchment to maintain production in a few key sectors, massive closure of local enterprises was effected through cutting off material supplies. Conversely, availability of substantial amounts of materials outside of plan channels renders state control much less effective. In a 1983 article in the Party journal *Hongqi*, Director of State Materials Bureau Li Kaixin complained that it is impossible to control local investment because of "insufficient material resources under state control" and the "excessive dispersion of materials" —at present, central government allocation of the three key construction materials of cement, steel and lumber accounts for only 25, 53, and 57 percent of total output, respectively. With large quantities of these materials circulating outside plan channels, it is much more difficult to shut down unwanted local projects. Instead, the present situation is said to be one where, "with money, materials can be obtained" (*youqian jiu youwu*).³

Unrestrained local investment is, in fact, one of the key problems that has plagued the Chinese economy since the early 1970s. In the post-Mao period, despite the government's repeated calls to curtail investment and rationalize production, local investment has continued to expand. In his report to the Sixth National People's Congress in June 1983, Vice-Premier Yao Yilin discussed some major problems "calling for attention and prompt solution." Chief among them was excessive investment in fixed assets, which exceeded the plan by 11 billion yuan in 1982. According to his analysis, "The main reasons for this were that capital construction investment in the form of funds collected by localities, departments, and enterprises themselves and investment financed by domestic bank loans exceeded the plan. The over-extension of capital construction and excessively decentralized use of the investment funds meant that some planned projects had to give way to those outside the plan and key projects to ordinary ones. Energy and transport projects, in particular, were adversely affected by a lack of funds, materials, and construction workers."⁴ This difficulty stems in large part from the progressive loss of central control over material and financial resources that began in the 1960s. This has occurred largely through the growth of local industries during the Cultural Revolution, which put control of very substantial amounts of resources in local hands, strengthening local governments vis-à-vis Beijing—both because local production came to account for growing portions of total output, and because the central government had to make a number of concessions in the material allocation system to encourage local investment, in accordance with the policy of self-reliance and self-sufficiency.

The policy of self-reliant industrialization at the local level had a fundamental impact on China's economic system—by the end of the Cultural Revolution, the Chinese economy was very much decentralized, with the balance of power between center and localities having shifted significantly toward the latter. This constitutes the major difference between the post-Leap retrenchment and the post-Mao readjustment of the late 1970s and early 1980s: In the post-Leap period, the central government was able quickly to reassert control over the economy, but in recent years it has been virtually powerless to do so. Significantly, the stated objective of returning some 10,000 key enterprises and over 600 producers' goods to direct ministerial supervision had not yet been implemented, even by mid-1983.[5] Having relinquished control of so many financial and material resources, the central government found its ability to influence local decision-making severely restricted: Withholding financial and material assistance from unwanted local projects can no longer ensure their termination. In this context, the state's ability to implement reforms of the economic system is necessarily limited by its ability to win acceptance of reforms from local authorities.

Throughout this chapter, *local* will refer to provincial level and below. However, *local industry* will include only enterprises built primarily with local finance; it excludes the approximately 10,000 enterprises decentralized to local control during the Cultural Revolution. For example, Anshan Iron and Steel is not a local enterprise under this definition, even though it is nominally under Liaoning provincial administration. This division is based on control over enterprise revenues and output. As will be shown below, this constitutes an important distinction, because local enterprises operate under constraints and incentives that are substantially different from those applying to enterprises under central control. Market reforms since 1978 have significantly loosened constraints on local industry and improved their incentives. As a result, local industry has responded vigorously to reform stimuli, but often in ways that run counter to reform objectives. In addition, the existence of a large local sector operating under quasi-market conditions exacerbates the unequal distribution of benefits and costs accruing to enterprises, significantly distorting the intended incentives of reforms.

THE MATERIAL ALLOCATION SYSTEM
AND LOCAL AUTONOMY

In the 1950s, as part of its planning model, China adopted the Soviet material-allocation system. Under this system, production materials are divided into three categories. Category I comprises the widely used, key materials under direct allocation by the State Planning Commission (*tongguan*). Specialized key materials under Category II are allocated by central ministries (*buguan*). Category III materials are under local allocation (*diguan*).[6] Materials flow downward to production enterprises according to their level of administration. That is, a centrally administered enterprise receives supplies from central government or ministerial departments, a provincial enterprise receives supplies from provincial departments, and so on. Category I and II materials distributed by local authorities are allocated by higher levels. (This flow pattern is depicted in Figure 1.) The pattern was somewhat altered when later decentralizations blurred the lines of enterprise supervision (discussed below). A second tract was eventually added whereby selected key enterprises were placed under "direct supply" (*zhigong*), which advanced them to the next higher level of supply: Some provincial enterprises were directly supplied by central ministries, some county enterprises were directly supplied by provincial departments, and so on.

Control of key producer and investment goods is an important instrument for directing resource allocation in the economy. Through the planned allocation of these vital materials, the state can determine what to produce, how to produce, and where to produce. Both the number of commodities under unified allocation and the number of enterprises under direct central government supervision fluctuate with swings of the centralization-decentralization pendulum (see Table 24). During the First Five-Year Plan period, both numbers rose as the economy was gradually brought under central planning. A peak was reached in 1957, with 532 commodities and 9,300 enterprises under central control. Materials under state allocation accounted for 70-90 percent of total available.[7] The first round of decentralization came in 1958 and was dramatic but brief:

FIGURE 1 Planned Material Flows

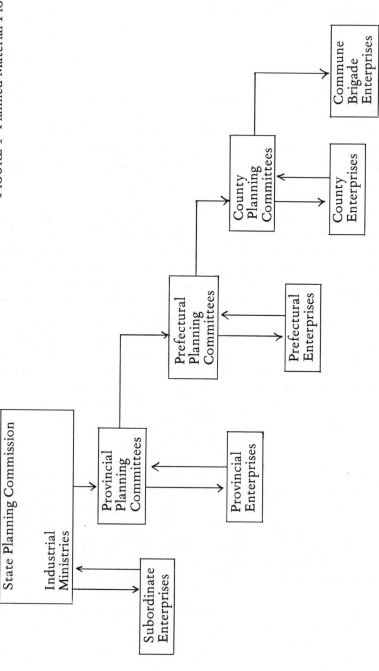

TABLE 24 Number of Producer Goods under State Allocation, 1953–1981

	Number of Goods in Categories I & II	Number of Enterprises under Central Control
1953	227	2,800+
1957	532	9,300+
1958	132	1,200
1960	432	2,000+
1963	522	10,000+
1964	592	NA
1965	579	
1966	579	
1971	217	
1972	226	
1978	210	
1979	64	
1980	64	
1981	67	

Sources: Liu Guoguang and Wang Ruisun, *Zhongguo de jingji tizhi gaige* (The reform of China's economic system, Beijing, 1982), pp. 4–8.
1960, 1963, 1965: Li Jingwen, "Problems Concerning the Nature and Management of the Circulation of Producers' Goods," in Liu Guoguang, ed., *Guomin jingji guanli tizhi gaige de ruogan lilun wenti* (Some theoretical problems of the reform of the national economic system; Beijing, 1981), pp. 203–204.
1964: Zhang Shushan, "Some Views about Reforms in Commodity Controls," *JJGL* 8:15–18 (1979); translated in JPRS, *Economic Affairs*, 34:16.
1980: Sun Xuewen, "During the Readjustment Period Reform Must Still Be Implemented," *JJGL* 2:7(1981).
1981: Interview, Beijing, May 1982.

87 percent of the enterprises were decentralized to local (mostly provincial) control, and the number of commodities under unified allocation was reduced by 75 percent. More important, according to one source, the system was changed from unified allocation at the national level to coordination at the provincial level. The central government set targets only for inter-provincial transfers of key materials, leaving the actual production, supply, and distribution plans to be formulated by provincial organs.[8] The extremely ambitious development programs of the Great Leap Forward quickly led to coordination problems, and, by the second half of 1959, central

ministries had begun to reassert control over material supplies. Centralization was restored to the 1957 level by 1963. With the post-Leap recovery well underway, planners moved once again to relax their grip on material allocations and to give localities more flexibility. This time, decentralization took a much more protracted form that included several stages and continued through the Cultural Revolution period. During 1953–1965, the central government began to allocate some portions of Category I and II materials for discretionary local use.[9] This was followed in 1966 by the transfer of almost all output from local small-scale industries to local allocation under the slogan "Whoever builds and manages the enterprise has the use of its output" (*shuijian, shuiguan, shuiyong*). With the development of the "five small industries" during the Cultural Revolution, local production came to include key producer goods such as iron and steel, cement, chemical fertilizer, coal, and farm machinery.[10]

Past scholarship has focused exclusively on the number of commodities under unified allocation as an index of centralized control.[11] However, the 1960s policies had in fact ended the state's allocative monopoly over Category I and II commodities. Indeed, during the 1971–1978 period, even though the number of commodities in these categories remained unchanged at about 210, material allocation had become increasingly regionalized with the growth of local industries, whose output was subject primarily to local allocation. This cessation of "unified allocation" of Category I and II commodities by the State Planning Commission and central ministries fundamentally changed the material supply system: Rather than dividing up control of materials according to their centrality or importance in production, allocation rights to materials became attached to ownership of the production unit. Coal produced in mines under central control came under central allocation; that produced in provincial mines came under provincial allocation, and so on.[12]

A distinction must be made between nominal and real control, however. When a large number of central enterprises were decentralized to local control in 1970, the central government did not in fact relinquish all rights to their output. Rather, to stimulate local initiative in operating these enterprises, localities were allowed to retain a portion of their output. In addition, they were allowed to add a supplementary output plan (called a *qichengshu*, an expected target) to the enterprises' mandatory target (*bichengshu*) assigned by the supervisory ministry. Even the lines of enterprise supervision became entangled: In 1975, for example, because of the "inability of many localities to run large-scale industries," over half the "decentralized" enterprises were directly supervised by ministries "on behalf of" the localities (*daiguan*). In these cases, assignment of production and investment targets, along with material supplies, continued to come from ministries, although other matters such as personnel often fell under local control.[13] Paradoxically, the formal decentralization of enterprises in 1970 had transferred relatively little control to lower levels, while real decentralization was taking place through the vehicle of local industrialization. The cumulative effect was that, by 1978, portions of total output of Category I and II commodities under local allocation included 46 percent of coal, 20 percent of steel, 64 percent of cement, 65 percent of machine tools, and so on (see Table 25).

In view of the declining portion of key materials remaining under central control by the end of the Cultural Revolution, it is not surprising that the state found it increasingly difficult to curb local industrial expansion. Two questions deserve exploration: How did it happen? What impact does this have on resource allocation?

Donnithorne, Perkins, and others have noted that backwardness and the great range of production techniques in use simultaneously in Chinese industry make the economy extremely difficult to bring under central planning.[14] Even at the peaks of centralized control in 1957–1958 and 1964–1965, unified allocation had been extended to only some 500 materials in China, compared with the Soviet Union, where 760 "funded" commodities were managed by Gosplan, along with some 5,000 centrally planned materials allocated by ministries.[15] As for the Cultural Revolution period, it seems clear that the further devolution of control over material allocation was the inevitable outcome of industrial policies pursued during that period.

TABLE 25 Local Control of Category I and II Materials, 1978

	Materials under Local Allocation (% of total)	Central Component (residual)	Central Component Turned Over for Local Distribution
Coal	46	54	66.6
Steel	20	80	28
Lumber	19	81	22.7
Cement	64	36	—
Machine Tools	65	35	32.4
Trucks	25	75	32.8
Nonferrous Metals	—	—	48
Pig Iron	—	—	35.9

Sources: Yu Xiaogu, "Implement the Merger of Planned and Market Adjustment to Liven up Material Circulation," *JJGL* 2: 4(1980).
Zhang Shushan, "Some Views about Reforms in Commodity Controls," *JJGL* 8: 15-18(1979), translated in JPRS, *Economic Affairs* 34: 16.

Rapid development of local industries resulted from a confluence of policies: The regional-self-reliance policy called for provinces and even prefectures and counties to build "relatively complete and independent" industrial systems; the defense strategy called for building industries widely dispersed throughout the interior provinces; and the technological policy of "walking on two legs" called for the development of small-scale production facilities alongside large ones. All these gave rise to a proliferation of medium- and small-scale enterprises. During 1970-1978, the number of enterprises rose from 195,000 to 348,000, many in the "five small industries."[16]

The impossibility of incorporating the large number of small enterprises into state material allocation plans necessitated a change in the planning structure. On the one hand, increasing portions of materials were turned over to local governments for distribution, thus transferring downward the responsibility for coordinating material supplies. According to one source, "Since 1970 many measures have been adopted to enlarge the authority of localities in allocating goods under unified distribution. The percentages of such goods turned over to localities in accordance with the state plan (were in 1978) as follows: 28 percent of steel products, 35.9 percent of pig iron, 66.6 percent of coal, 22.7 percent of lumber, 48 percent of nonferrous metals, 32.8 percent of motor vehicles, and 32.4 percent of machine tools."[17] Another source reported that, during the Fourth Five-Year Plan, a "26-character principle" was implemented for the management of key materials such as coal, cement, and lumber, whereby only targets for interprovincial transfers were set by the state, leaving detailed balancing to local authorities.[18]

At the same time, local self-reliance was encouraged so as to supplement state resources. To provide the incentive and flexibility for localities to pursue self-reliant development, further concessions were made. In addition to the 1966 and 1970 measures discussed earlier, which gave localities control over output from local enterprises, the state had to tolerate an increasing amount of barter trade:

Along with an increase in distribution of resources by localities has come a more lively movement of goods; a cooperation in goods had developed among regions, and a gradual increase has taken place in the mutual supply of each other's needs. This sort of cooperation in the mutual exchange of goods under local control has become a rather active method for the exchange of goods among regions.[19]

It is difficult to estimate the magnitude of flows through these extra-plan channels, especially since many of them were in contravention of state regulations.[20] What is clear is that, as low-priority claimants in the hierarchical system of material allocations, local—especially sub-provincial—enterprises faced serious shortfalls in meeting production requirements during the 1970s, and had to turn to market channels for much of their supplies. Anecdotal accounts indicate that barter trade constituted a major source of supply in some localities. One report from Wuxi county remarked that machinery and equipment produced in the county were commonly used as "hard currency" (*yingtonghuo*) in "cooperation arrangements" to obtain most of the 50,000 tons of steel annually consumed in the county's commune and brigade enterprises. In addition, arrangements with Shanxi coal mines not only supplied enough coal for the county's industrial and household use, but also left a surplus which the county sent to a neighboring chemical-fertilizer plant to be "processed" (*jiagong*) into fertilizer.[21]

The hierarchical and regional material-allocation system that evolved during the Cultural Revolution greatly enhanced the role of local authorities in coordinating supply and marketing for enterprises within their administrative units.[22] Indeed, the speed of local development often hinged upon their entrepreneurial skills in ensuring smooth and growing supplies of needed inputs for local enterprises.[23] Popular slogans of the period, which urged localities to "develop self-reliantly rather than wait for handouts from the state" (*zili gengsheng, bukao guojia*), "build up industries by the 'snowballing' method," and "raise hens to lay eggs, hatch eggs to raise chickens," and so on, provided the ideological sanction for local authorities to pool resources within their administrative areas by actively intervening in local enterprise affairs and requisitioning enterprise output and funds.

Along with the growing control over production and distribution, during this period local governments also obtained increased financial autonomy. In 1966, rules were altered to allow local enterprises to retain depreciation funds. In 1972, the state began to allocate 2 yuan per capita to the provinces as discretionary funds for local budgets.[24] Together, these completed the necessary conditions for rapid local industrialization. Through the Third and Fourth Five-Year Plans, the locus of industrial expansion shifted increasingly to the local sector. In the iron and steel industry, for example, whereas local investment had accounted for only 8 percent of the total during the First Five-Year Plan, it had risen to 51.9 percent during the Fourth Five-Year Plan (1971-1975).[25] The effort to increase self-sufficiency in coal supplies in the seven provinces of Jiangnan also relied overwhelmingly on local investment: Of the 59 million tons produced there in 1979, 86 percent came from locally run coal mines.[26] By the end of the Cultural Revolution, local governments clearly emerged as powerful forces in the economic sphere.

IMPACT ON RESOURCE ALLOCATION

To the extent that local balancing and inter-regional trade helped mobilize resources for local development, the material-allocation system that evolved during the Cultural Revolution period contributed to the overall rate of growth. At the same time, it exacerbated the tendency—common to all planned economies—for hoarding and duplicative local production. By various accounts, rapid economic growth and the development of self-sufficient industrial systems were the primary objectives during the Cultural Revolution, for which development of infrastructure and social services were routinely sacrificed as local governments squeezed "nonproductive" expenditures to divert funds for "productive" investment in heavy industry.[27] Under conditions of excessive investment and growing shortages, when supplies from the state fell increasingly short of production requirements during the 1970s, many localities resorted to duplicating production facilities.[28] In an unusually detailed account, one writer from the State Capital Construction Commission explained how four separate and comprehensive iron and steel mills came to be built in the medium-sized city of Huangshi: Even though the city already had the Daye Iron and Steel Mill, a large-scale, centrally administered enterprise, the set-up was such that "all its profits were handed over to the state and the locality did not gain any benefits. Therefore, the Hubei provincial authorities built their own Xialu Steel Works in Huangshi." Likewise, the province kept all output produced in Xialu: "Steel was not available to the locality, and therefore the municipal government set up its own steel works, the Second Huangshi Steel Works, in the neighborhood of the national and provincial steel mills. Similarly, the municipal authorities did not care about the interests of the suburban counties . . . ." Since it was allocated only 300 tons of the 3,000 tons of pig iron needed annually, Daye county also built its own iron and steel mill. "In this way, redundant construction has occurred. The leading comrades of the municipal and county authorities told us that they had to operate their own iron and steel works even though they knew this would incur losses. The system 'forced people to build' these mills." Similar duplication prevailed in other industries in Huangshi: Aside from the Huaxin Cement Works, a large, central enterprise, there were 23 other cement plants operated by the provincial, municipal, county, and commune authorities, and even by the armed forces. As might be expected, the intense competition for investment funds, raw materials, equipment, and energy resources had a clearly adverse impact on efficiency: All plants operated below capacity, with incomplete sets of equipment. In the analysis of the report, "These are the inevitable consequences of the fact that the administrative bodies at various levels operate their own enterprises, are overly preoccupied with their own benefits and do not pay attention to the interests of various other parties."[29] In addition, this fragmentation necessitated more warehouses and inventories at each level, exacerbating shortages.

To alter this hierarchical and regionally fragmented material-allocations system, post-Mao reforms have tried to improve horizontal flows by increasing the role of the market. In 1979, the number of goods under unified allocation was reduced from 211 to 64.[30] In addition, even more of the remaining Category I and II materials were made available outside state allocation channels: State enterprises are now allowed to sell a portion of their above-plan output on the market. Enterprises and materials bureaux have been directed to sell off excess inventories, many restrictions on inter-regional trade were lifted, prices of many producers' goods were allowed to float, and numerous "materials exchange fairs" (*wuzi jiaoliuhui*) have been organized by local authorities.[31] As a result of these changes, central control over key materials was further reduced (see Table 26), and the market for producers' goods has greatly enlarged its scope. According to the Director of the State Materials Bureau, Li Kaixin, by early 1983, materials traded outside plan channels were accounting for some 20–30 percent of the total at the provincial level, 30–40 percent at the prefectural or municipal level, and over 50 percent at the county level.[32] In Hebei province, 58.3 percent of steel products handled by the provincial materials departments reportedly came from extra-plan channels in 1982.[33] In Liaoning province, since 1979, steel, lumber, and cement from extra-plan sources have supplied 30, 50, and

TABLE 26 Local Control of Category I and II Materials
(as Percent of Total Output)

	1965	1980	1982
Coal	25	46	49
Steel	5	42	47
Lumber	37	18	43
Cement	29	71	75
Nonferrous Metals	NA	36	NA

Source: Li Kaixin, Director of State Materials Bureau, "Management of China's Materials and Resources," *1981 Jingji nianjian* (1981 almanac of China's economy), IV, 124.
Li Kaixin, "Concentrate Materials to Ensure Keypoint Construction," *Hongqi* 17: 16 (1983).

70 percent of the total allocated quantities at the provincial, city, and county levels, respectively.[34]

Rather than marking a fundamental break from past practices, the post-Mao changes merely accelerated the erosion of central control over material allocations. The key differences between the Cultural Revolution period and the post-Mao period are: (1) the volume of extra-plan trade has grown significantly; (2) a much greater portion of this trade is now conducted in the open; (3) there is greater price flexibility, and (4) there is a clearer delineation between plan and extra-plan materials, with restrictions on the movement of extra-plan materials virtually eliminated.

The Chinese industrial system has long been characterized by a dual structure, with a planned sector (large-scale enterprises at the central and sometimes provincial levels) accompanied by a quasi-market fringe (small-scale enterprises at the local level).[35] In the post-Mao period, the scope of the quasi-market fringe has expanded to include many more activities as well as enterprises formerly in the planned sector. For example, many machine-building enterprises had their state production assignments drastically reduced in 1979 and were told to go out to solicit orders. For the local, small-scale enterprises, which were only partially incorporated in state plans during the Cultural Revolution, the reforms lifted many restrictions on supply and marketing activities. However, since the reforms did not fundamentally alter the relationship between local governments and their subordinate enterprises, they leave the quasi-market "piece" of the economy firmly under local government control, with local enterprises more dependent than ever on local planning and commercial networks.[36]

This encystment of local economies is exacerbated by recent financial reforms. To improve incentives for better local supervision of enterprises, in 1980 the state introduced revenue-sharing schemes that allowed most provinces to retain all or a fixed portion of profits from local enterprises.[37] In many provinces, this system was extended downward through revenue-sharing arrangements with counties. Rather than passing along profits and losses of their enterprises to the state budget, local governments are now more or less independent accounting units responsible for the financial health of these enterprises. Although rapid growth is no longer enshrined as the principal objective, local governments are under pressure to "repay old debts" by building urgently needed urban infrastructure, especially housing, which is financed largely out of local revenues. The return of 20 million educated youths from the countryside since 1979 has also created an enormous burden on the local authorities to find jobs for them in the urban areas, including county towns. The effect of these changes has been to transform local governments into much more autonomous economic agents, with a strong incentive to maximize revenues and employment.

LOCAL GOVERNMENT AS ECONOMIC AGENTS

In some respects, these increased market activities and the new incentive system have helped improve resource flows in the Chinese economy. Freeing large numbers of machinery and electrical products from state allocation has facilitated the shift from heavy industry to increased consumer-goods production. Chief beneficiaries are the small, local plants which can now purchase equipment and other key materials formerly allocated only to larger production units at higher levels. The massive entry of these small plants into consumer-goods production has helped improve supplies. Allowing extra-plan producers' goods to be traded at "negotiated" prices has also, to some extent, helped redirect resources toward areas where they are most needed. With market prices bid up to levels much closer to scarcity values, resources are, at the margin, attracted to or withdrawn from sectors according to supply and demand conditions. For example, the very high negotiated price of coal (50-80 yuan per ton, compared with 20-30 yuan under state allocation) has accelerated growth of coal production in some areas.[38] Shanxi province has nearly doubled its output from 1976 to 1981.[39] Similarly, the post-Mao construction boom has sent negotiated prices of construction materials soaring, which have, in turn, attracted resources to that sector.

With improved financial incentives, local governments are taking a more active interest in steering development along lines of comparative advantage, with Shanxi province stressing coal production, Yunnan and Guizhou speeding up exploitation of phosphorus ores, and so on. At the same time, many of the most inefficient and costly plants in the "five small industries" have been closed as local governments shed unwanted financial burdens. The vibrancy of local economies that has impressed recent visitors lends support to the view that current reforms have succeeded in tapping local initiative that had lain dormant under the more rigid system in operation during the Cultural Revolution.

Aside, however, from the problem of absorbing vigorous local development at a time when the state is trying to limit overall investment, some of this local activity is misguided. Faced with much greater availability of producer and investment goods in the market, and suddenly flush with investible funds made available by the recent fiscal decentralization, local governments have plunged into what they see as revenue- and employment-generating activities. Given a price structure that is still sending out many false signals, local revenue-maximization often led to duplication of production in the high-profit industries. Table 27 presents a list of profit rates for selected industries. With ideological and administrative restrictions to entry now lifted, some of these industries are extremely attractive to local investment. For example, the Shanghai Watch Factory produces wristwatches at a unit cost of 12 yuan. Until recently their watches sold for 120 yuan retail. Such large profit margins

TABLE 27 Profit and Industrial-Commercial Tax Rates for
Selected Industries[a]

(%)

	Profit Rates[b]	Tax Rates[c]
Watchmaking	61.1	40
Rubber processing	44.9	10–18
Bicycles	39.8	15
Dyestuffs	38.4	8
Daily utensils	30	—
Pharmaceuticals	33.1	5
Cotton textiles	32.3	5–18
Chemical fertilizer	1.4	3
Iron ore mining	1.6	5
Coal mining	2.1	8
Shipbuilding	2.8	—
Farm tools	3.1	3
Cement	4.4	15
Farm machinery	5.1	3

Notes: [a]Profit rates are probably based on production cost, although it is not specified in the source.

[b]Wang Zhenzhi and Wang Yongzhi, "Epilogue: Prices in China," in Lin Wei and Arnold Chao, eds., *China's Economic Reforms* (Philadelphia, University of Pennsylvania Press, 1982), p. 229.

[c]*Guojia shuishou* (State taxes), edited by the Teaching Material Editing Group for State Taxes (Beijing, 1982), pp. 57-62.

both attract new entrants and protect high-cost producers: Even with production costs that are two to three times that of the Shanghai plant, local watchmakers still make substantial profits at current prices.[40] In a number of industries, including watchmaking, bicycles, textiles, electrical appliances, and light industrial equipment, recent expansion has far outstripped demand as well as input supplies.

Some may argue that these are simply short-run, adjustment problems of the current market reforms. If allowed to work its usual magic, the competitive market will eliminate price distortions over time and drive out inefficient producers. Indeed, after an initial two or three years of inflation, by 1982 there was widespread price-cutting as competition sharpened in some consumer goods markets. In Tianjin, for example, municipal officials complained of the market being flooded with "outside" synthetic fabrics that were selling for 10-30 percent less than Tianjin's own products.[41] For wristwatches, competition had driven prices down to 30-60 yuan for some local brands.[42] In producers' goods, too, as the market has settled, increased competition in light industrial machinery is driving out some of the less efficient producers. By mid-1982, some of the commune and brigade factories that had rushed headlong into production of knitting machines were being squeezed out by larger, more sophisticated producers.[43] The nature of this market and the type of competition that can take place are, however, very much distorted by the active intervention of local governments, which manipulate the environment in a variety of ways to benefit local enterprises.

In this endeavor, they get some help from the state itself, in the form of contradictory components in the reform package. A prime example is the partial nature of price flexibility. At present, flexible prices are enjoyed mostly by local small plants, with the consent of local officials. Large-scale, state enterprises are allowed to sell only a portion of their above-plan output at negotiated prices. The bulk of their output must still be turned over to state commercial or materials bureaus at state prices. For products whose demand exceeds supply and whose negotiated prices are bid upward, the increased profits accrue disproportionately to small plants, attracting still more small-scale entrants to the industry. In coal mining, for example, local small mines and commune-run mines have cropped up everywhere to take advantage of negotiated prices of 50-80 yuan per ton, while state mines are still receiving only 20-30 yuan for the bulk of their output.[44] Similarly, even though small plate-glass plants have costs that are often twice those of larger plants, they are extremely profitable, since they can sell their output at three to four times the state price.[45] In one Guangdong county, commune-run cement plants were selling their output at 140 yuan per ton and reaping substantial profits, while county-run plants had to sell at the money-losing state price of 96 yuan.[46] The result is an "abnormal" situation where "whatever is under state control is sold at unified state allocation prices, which are fairly low, and whatever is not state-controlled is of inferior quality but is sold at high prices."[47]

Where supply exceeds demand and prices are falling, while small plants are cutting prices to compete, large plants operating with fixed state prices again find their hands tied. For the Shanghai Watch Factory, for example, even though they were allowed in 1982 to cut prices to 80 yuan, they still found themselves faced with a shrinking market as local watchmakers cut their prices even further.[48] With rules of the game so biased against the large-scale, state-run enterprises, it is they who often turn out to be the victims of competition. Lacking the ability to cut prices to match their competitors, Shanghai firms were finding it increasingly difficult to sell even formerly coveted products such as transistor radios, electric fans, cigarettes, candies, textiles, and so on.[49] Efforts to curb either unauthorized price changes or production have been predictably unsuccessful. Press reports complain that restrictions on price-cutting or price increases are routinely circumvented as enterprises downgrade or upgrade the product in question. Similarly, when attempts are made to impose output ceilings to cut back on production of overstocked products, only large-scale, state enterprises are affected. Local small plants continue to do as they please under the protective wings of local authorities.[50]

With local governments acting as economic agents, market signals are also complicated by taxes and the fiscal arrangement. In a 1983 article, Xue Muqiao explained that, in the past, high tax rates had been used as a lever to control both consumption and production in the cigarette and wine-making industries, by keeping prices high and profits low. In recent years, however, small cigarette plants and wineries have blossomed everywhere. During 1978–1980. tax exemptions for small plants run by counties, communes, and brigades made them extremely profitable: Although originally intended only to cover profit taxes, the concessions were extended to include the 40–60 percent industrial-commercial (turnover) tax in local implementation. When these exemptions were rescinded in 1980, however, fiscal decentralization had transferred cigarette and liquor taxes to local revenue income.[51] "With 'cooking in separate kitchens' extended down to the counties, taxes paid by county enterprises go into county revenues, and taxes paid by provincial enterprises go into provincial revenues. So, for cigarette and liquor taxes, the taxpayer is the county, and the tax collector is also the county. In reality, this is like having no taxes at all, so that taxes can no longer be used as an economic lever."[52] In fact, for purposes of revenue-maximization, taxes and profits are indistinguishable to local governments, and the problem of "irrational" prices is magnified by "irrational" taxes to distort resource allocation. Even where competition drives down prices, or where high production costs eliminate profits, the undertaking remains lucrative as long as high taxes ensure increased net revenues to local coffers. Table 28 provides a list of industries with high industrial-commercial tax rates which have attracted a great deal of local investment in recent years.

TABLE 28 Selected Industries with High Industrial-Commercial Tax Rates

	% of price
Tobacco: Grade A and B cigarettes	66
Grade C cigarettes	63
Grade D cigarettes	60
Grade F cigarettes and shredded tobacco	40
cigars	55
Wine-Making: Grain-based white and yellow wines	60
beer, potato wines, local sweet wines	40
fruit wine, medicinal wine	15
Sugar: Machine-made cane sugar	40
hand-made cane sugar, beet sugar	30
Miscellaneous:	
clocks, cameras, and electric fans	25
sewing machines	10

Source: *Guojia shuishou* (State taxes), pp. 61-62.

In terms of center-local relations, the more troublesome fact is that profit-retention schemes and fiscal decentralization have combined to sharpen the differentiation between economic activities that take place inside and outside an administrative unit's borders, which will greatly exacerbate the "cellular" characteristic of the Chinese economy. The unequal distribution of profits between producers and processors of raw materials under the current structure of prices had long been a source of conflict among regions. Post-Mao reforms that enhanced local power have exacerbated problems of localism and increased resistance to inter-regional transfers.[53] Indeed, the recurring problems of duplicative production are the direct outcomes of local enterprises and governments trying to capture a larger share of profits and taxes. For example, given the high profit and tax rates for cigarette manufacture, local officials would prefer to see tobacco processed within their administrative boundaries, rather than passing them along to Shanghai or Tianjin. With cigarette tax revenues accounting for some 30 percent of its total revenue income under the present fiscal system, Henan province has a very strong incentive to fight to retain more tobacco for local processing.[54] The same applies to an even greater extent to subprovincial units, which are often greatly dependent on specific taxes for revenue income.

Ironically, the growing problems of effecting planned inter-regional transfers of resources at state prices are accompanied by flourishing inter-regional trade in the reform period, mostly in the form of bilateral "cooperation agreements" between localities. For example, one suburban county in Shanghai signed a contract with Fengyang county in Anhui to provide technical assistance in exchange for deliveries of 10,000 to 15,000 tons of fodder grain per year. Other counties in the municipality agreed to invest a total of 40 million yuan in coal-mining development in Shanxi province in exchange for future deliveries.[55] In an era when economic incentives are emphasized, there is the phenomenon in China where inter-regional trade is increasingly conducted at "border" prices that exceed prices within the locality. A 1981 survey of 14 counties in Shanxi found that, while coal sold to local users fetched prices averaging 8.04 yuan per ton for industry and 6.77 yuan for household use, that sold "outside" (*waixiao*) earned over 27 yuan per ton.[56] For such bilateral trade between localities, even when the agreements are denominated in state prices, these goods and services are effectively being traded at market or quasi-market prices. With diminishing state control, localities are increasingly having to be induced to give up their resources by offers of higher prices.

In consumer-goods markets, regionalism is often manifested in protectionism against competition from outside. In their eagerness to expand their revenue base and to create jobs, coupled with inexperience, many local cadres have over-invested in consumer-goods industries in the past few years. With increased competition in the domestic market and an unstable export market, many enterprises have run into difficulties. Because local revenues are so dependent on enterprise taxes and profits, these local cadres must often resort to protectionist tactics to shore up revenues as well as to cover up mistakes in investment decisions. With near-complete control over the commercial network and retail outlets, local authorities can set up a variety of obstacles to "imports." Press reports are filled with complaints of local officials refusing to purchase or even display competitors' products, unreasonably raising the prices of "imports" to make them uncompetitive, and so on.[57] Until the commercial system can be wrested from local government control and reorganized along economic lines, there will always be the temptation for local authorities to use it to shield local enterprises against competitive pressure. In such cases, the efficiency gains from a competitive market will not be realized. Instead, Chinese industry will continue to be characterized by monopolistic competition, with numerous high-cost firms operating with excess capacity and serving small, fragmented markets.

CONCLUSION

Even though decentralized decision-making in the past few years has brought some improvements, serious problems loom on the horizon. On the macroeconomic level, greatest increases in efficiency are likely to come from efforts to correct the structure and composition of industry: to get a better fit between supply and demand, to raise the technical level, to reduce duplication, and to realize specialization and economies of scale. These cannot be effected until investment is brought under control. In the post-Mao period, since decentralization has exacerbated the twin problems of over-investment and duplication, it has contributed to the competition for supplies, shortages, and bottlenecks, which are in turn thwarting efforts at instilling cost-consciousness in enterprise operations. In addition, the proliferation of local projects at the expense of large-scale state investment biases the choice of technique toward small-scale production units, which may well have adverse long-term implications for efficiency. For example, due to the high energy consumption in many small cement plants, the Ministry of Building Materials called in 1978 for a three-year moratorium on building new ones.[58] This has been widely ignored. Through 1982, there had been a net addition of 600 enterprises in the small-scale sector, which raised its share of total output from 64 to 72.5 percent in one of the fastest growing industries in the post-Mao period.[59] Nanjing municipality provides a typical example: During 1978-1981, the city built 17 new plants in response to new demands generated by the post-Mao construction boom, raising the city's annual production capacity from 100,000 to 400,000 tons of cement.[60]

The present situation is inherently unstable in two important respects. With the locus of power shifting during the reform period, there is intense competition for control among administrative units—between center and province, between province and county, among provinces, among counties, and so forth. Having gained control over vast amounts of resources, local governments were, by the end of the Cultural Revolution, in a powerful position to push reforms along the lines of further decentralization and to resist efforts to reassert central control. As long as reforms continue to enhance their power, local governments constitute a major force behind their momentum. (The constraint placed by this power structure on the configuration of post-Mao reforms is masked by the widespread support within the central leadership itself for market reforms. In embarking on the current reforms, the Chinese leaders were undoubtedly motivated by the feeling that central planning in the previous decade had not been effective in allocating resources efficiently to produce goods that were needed, and that adjustments necessary to correct these problems could be better handled in a decentralized way by localities and enterprises.) Through 1982, reforms had continued to shift the balance of power toward local governments by transferring more financial and material resources to their control. With each successive budgetary crisis, the central government seems to be conceding further control in negotiations with provinces (see Barry Naughton's chapter in this volume).

There is also the unstable balance between plan and market, with high market prices bidding away resources intended for plan channels. Underfulfillment of state plans is a growing problem at all levels, with local authorities diverting resources intended for state projects, enterprises making unauthorized changes in the output mix to maximize profits, and so on. Because of widespread underfulfillment of planned deliveries, the state was reportedly forced in 1983 to buy some 23 million tons of coal and 4 million tons of cement from extra-plan channels at negotiated prices, in addition to importing steel products from abroad, in order to ensure adequate supplies to key-point construction projects.[61] While the efficiency implications of such actions may be debatable, their costs added 600 million yuan to state expenditures, further straining the financial resources of the government and perhaps hastening the reforms' demise.

The present trend toward progressive weakening of central control also implies potential welfare losses where projects (especially in energy exploitation and transportation) embody external economies that require higher-level government intervention.[62] In addition, the decline of central power means the decline, at least potentially, of the government's ability to effect income transfers across administrative units. The inability to redistribute resources and income across administrative boundaries that had plagued rural development over the past three decades may now be extended to other sectors of the economy as well. With each locality jealously guarding its resources, interregional trade is increasingly commanding premium prices. A decentralized, market system that allows residents of richly endowed regions to be the sole beneficiaries of economic rents accruing to those resources represents a great leap backward in the transition toward a solidary society.

Many of the current problems stem from the piecemeal nature of the reforms and the remaining rigidities in the system, such as the different incentives that apply to different enterprises, a market that handles only a portion of production and sales, and prices that are only partially flexible. The multi-tiered price structure that has developed is creating substantial pressure for a generalized price reform, which is said to be forthcoming. A price reform that adjusts the present administrative prices to reflect more closely relative scarcities will help alleviate some of the problems and improve competition among enterprises. Price reform will only go part way, however, toward eliminating distortions caused by divergent profitability across industries. As long as local governments remain the primary economic agents in making investment and production decisions, there will be continuing administrative intervention to fragment the market along regional boundaries. The incentive structure will also remain distorted by the dissimilar industrial-commercial tax rates across industries. The tax-for-profit scheme and the surtax on extra-budgetary funds introduced in 1983 may be steps in the right direction toward reducing the influence of local governments. The most fundamental problems in the system, then, are those underlying the relationships between the center and localities, and between the state and the enterprises. How to give greater autonomy to the microeconomic units and induce them to interact in ways that produce outcomes consistent with social objectives is a problem that has not yet found its solution in any society.

Notes
Index

NOTES

Introduction: The Political Economy of Reform in Post-Mao China: Causes, Content, and Consequences, by Elizabeth J. Perry and Christine Wong

1. State Statistical Bureau, *1981 Tongji nianjian* (1981 statistical yearbook; Beijing, 1982), p. 225.
2. Thomas B. Wiens, "Technological Change," in Randolph Barker and Radha Sinha, eds., *The Chinese Agricultural Economy* (Boulder, Westview Press, 1982), pp. 99-120.
3. Ma Hong and Sun Shangqing, eds., *Zhongguo jingji jiegou wenti yanjiu* (Research on the problems of China's economic structure; Beijing, 1981), p. 3.
4. *People's Daily*, 30 March 1981, p. 5.
5. *JJGL* 6:9 (1979).
6. State Statistical Bureau, "Communiqué on the Fulfillment of China's 1979 National Economic Plan," *Beijing Review*, 6 July 1979, pp. 37-41.
7. See, for example, Andrew Walder, "Industrial Reform in China: The Human Dimension," in Ronald A. Morse, ed., *The Limits of Reform in China* (Boulder, Westview Press, 1983).
8. Samuel Huntington, *Political Order in Changing Societies* (New Haven, Yale University Press, 1968), p. 345.

9. For discussion of policy shifts, see G. William Skinner and Edwin A. Winckler, "Compliance Succession in Rural Communist China: A Cyclical Theory," in Amitai Etzioni, ed., *Complex Organizations: A Sociological Reader* (New York, Holt, Rinehart, and Winston, 1969); and Paul Hiniker and Jolanda Perlstein, "Alternation of Charismatic and Bureaucratic Styles of Leadership in Post-revolutionary China," *Comparative Political Studies* 10 (1978).
10. See, for example, George Breslauer, "Political Succession and the Soviet Policy Agenda," *Problems in Communism* (May-June 1980); Valerie Bunce, *Do New Leaders Make a Difference? Executive Succession and Public Policy under Capitalism and Socialism* (Princeton, Princeton University Press, 1981); and Joseph W. Esherick and Elizabeth J. Perry, "Leadership Succession in the People's Republic of China: Crisis or Opportunity?" *Studies in Comparative Communism* (Fall 1983).
11. As the Party theoretical journal, *Hongqi*, editorialized in February 1977, "It is good to take care of the well-being of the masses." The editorial noted that economic growth must apply to consumer goods such as bicycles, cotton, and sugar.
12. Bunce, pp. 12-38.
13. William Byrd, "Enterprise-level Reform in Chinese State-owned Industry," *American Economic Review*, May 1983.
14. In contrast, an early retirement program for urban state-owned enterprise workers—which permitted retirees to be replaced by their children—was embraced enthusiastically. Indeed, many factories were complaining about its hemorrhage effect on skilled workers, who were anxious to pass their prize jobs along to sons or daughters before the policy was revoked. In such cases, the retired workers could sometimes even continue to work as consultants, thus drawing both the retirement pension and the consulting income. This is a process that often selects the unfittest: Jobs are frequently passed along to the child least capable of fending for himself.
15. *Xinhua*, Beijing, 20 April 1984, FBIS 24 April, K16-17.
16. *Beijing Review*, 9 April 1984, pp. 4-5.
17. Huntington, p. 347.
18. Nicholas R. Lardy, "Agricultural Prices in China," World Bank Staff Working Paper no. 606, September 1982.
19. *1983 Tongji nianjian*, p. 235.
20. *Caimao jingji* (Finance and trade economics) 9:50 (1982).
21. William Hsiao, "Transformation of Health Care in China," *New England Journal of Medicine*, 5 April 1984, p. 932.
22. Ibid., p. 934.

Chapter 1. Socialist Agriculture is Dead; Long Live Socialist Agriculture! by Kathleen Hartford

The research on which this paper is based was made possible by a postdoctoral research fellowship from the Harvard Business School in 1981-1983. My Hong Kong interviews in June 1983 were facilitated by the Universities Service Centre, and indispensably aided by my research assistants, who have requested anonymity. The research and writing received invaluable help from James Austin, Thomas P. Bernstein, William Byrd, Anita Chan, Paul Cohen, Michael Oksenberg, Suzanne Pepper, Dwight Perkins, Elizabeth J. Perry, Michael Reich, Peter Timmer, Benedict Stavis, Jonathan Unger, Karl-Eugen Wädekin, Christine Wong, David Zweig, and many colleagues at the Fairbank Center for East Asian Research at Harvard University.

1. For basic introductions to the Maoist mode, see: Byung-joon Ahn, "The Political Economy of the People's Commune in China: Changes and Continuities," *Journal of Asian Studies* 34.3:633-658 (May 1975); Frederick W. Crook, "The Commune System in the People's Republic of China, 1963-1974," in *China: A Reassessment of the Economy*, A Compendium of Papers Submitted to the Joint Economic Committee, Congress of the United States (Washington: U.S. Government Printing Office, 1975), pp. 366-410.
2. See, for example, "Zhonggong zhongyang guanyu jiakuai nongye fazhan ruogan wenti di jueding" (Decision of the CCP Central Committee on some questions concerning accelerating agricultural development), *Zhongguo nongye nianjian 1980* (China agricultural yearbook, 1980; Beijing, Nongye chubanshe, 1980), pp. 56-57; Xue Xin, "Scientific Socialism or Agrarian Socialism?" *Social Sciences in China* 1:85-88 (March 1982).
3. Da Fengquan, "Nongye lianchan chengbao zerenzhi tantao" (An exploration of the agricultural responsibility contracts linked to output), *Nongye jingji wenti* 1:42-43 (1983); Su Xing, "Zerenzhi yu nongcun jiti suoyouzhi jingji di fazhan" (The responsibility systems and the development of the collective-ownership economy), *JJYJ* 11:3-4 (1982).
4. Liu Xumao, "Woguo nongcun xianxing di jizhong zhuyao shengchan zerenzhi jianjie" (A brief introduction to several principal systems of production responsibility now practiced in our country's villages), *JJGL* 9:12-14 (September 1981); Wu Guanming, "Qingpu xian Linjiacao dadui shixing zhuanye chengbao jingji xiaoyi xianchu" (The economic results of Qingpu county's Linjiacao Brigade carrying out specialized contracts are obvious), *JJGL* 7:69-72 (1982).
5. Liu Xumao, "Woguo nongcun," p. 13.
6. Information on this team comes from two 3-hour interviews with informant GGR, Hong Kong, June 1983.
7. Based on four 2-hour interviews with informant GQW, Hong Kong, June 1983.

8. Interviews with YNU, QIU, GGR, YYG (all indicated unspecified-term or long-term contracts without change), and GQW, GGO (both indicated adjustments could be made after a year), Hong Kong, June 1983.
9. *People's Daily*, 25 August 1983, p. 2.
10. CCPCC, "Concerning Further Strengthening and Perfecting the Systems of Job Responsibility for Agricultural Production (September 1980)," FBIS 20 May 1981, K8-11; "Minutes" of CCP 1981 Rural Work Conference, FBIS 7 April 1982, K1-3 passim.
11. *People's Daily*, 3 April 1982, p. 1.
12. Su Xing, "Zerenzhi yu nongcun jiti suoyouzhi," p. 6; *Beijing Review*, 5 September 1983, p. 6; JPRS, *China Report: Agriculture* 276:5 (24 October 1983).
13. David Zweig, "Opposition to Change in Rural China: The System of Responsibility and People's Communes," *Asian Survey* (July 1983), pp. 879-890; see also Jack Gray and Maisie Gray, "China's New Agricultural Revolution," in Stephan Feuchtwang and Athar Hussain, eds., *The Chinese Economic Reforms* (London, Croom Helm, 1983), pp. 178-179.
14. Interview with GGR, cited above.
15. Two 3-hour interviews with informant QIU, Hong Kong, June 1983.
16. "The 6th Five-Year Plan," *Beijing Review*, 12 October 1981, p. 3: "Readjustment of Agricultural Structure Pushed," JPRS, *China Report: Agriculture* 154:10 (23 July 1981); Zhou Hui, "Actively Guide Rural Surplus Workforce to Devote Efforts to 'The Land,'" JPRS, *China Report: Agriculture* 226:25 (16 September 1982).
17. State Agricultural Commission, "Report on the Active Development of a Diversified Rural Economy," JPRS, *China Report: Agriculture* 159:7 (17 August 1981); *Beijing Review*, 29 June 1981, pp. 3-4; Zheng Jianwei and Feng Jian, "Always Keep in Mind the 800 Million Peasants—A Report on Zhongnanhai," FBIS 20 May 1981, K9; "Fanrong nongcun jingji di zhanlue cuoshi" (A strategic measure for enlivening the rural economy), *People's Daily*, 2 May 1981, p. 4.
18. JPRS, *China Report: Agriculture* 273:2 (23 September 1983); and "*Renmin ribao* hails peasants entering into business," JPRS, *China Report: Agriculture* 249:43 (16 February 1983).
19. Interview with GGR, cited above.
20. Two 3-hour interviews with informant GNW, Hong Kong, June 1983.
21. "New Changes Take Place in Urban and Rural Country Fair Trade Nationwide," JPRS, *China Report: Agriculture* 261:24 (15 June 1983); "Communiqué on Fulfillment of China's 1982 National Economic Plan," *Beijing Review*, 9 May 1983, p. viii.
22. "CCP Document No. 1 on Rural Economic Policies," FBIS 13 April 1983, K6-7.
23. *Almanac of China's Economy 1982*, JPRS, *China Report: Economics*

370:194 (8 August 1983); *JJNJ 1983* (China economic yearbook 1983; Beijing, 1983), p. III–42.
24. *Beijing Review,* 5 September 1983, pp. 6-7; *Beijing Review,* 27 February 1984, p. 4; "'Specialized' Families Utilize Labor, Resources," JPRS, *China Report: Agriculture* 253:126 (6 April 1983).
25. JPRS, *China Report: Agriculture* 275:60-66 (17 October 1983) and 279:63-64 (17 November 1983).
26. Yu Guoyan, "Discussion of Problems in Specialized Rural Households," JPRS, *China Report: Agriculture* 245:26-28 (7 January 1983); and Ma Encheng, "Characteristics and Trends of Development of Rural Specialized Households," JPRS, *China Report: Agriculture* 254:103 (18 April 1983).
27. Ibid., p. 34; Yu Zuyao, "Contracting Agricultural Output Quotas to Individual Households Is a Prelude to Economic Reform in China," JPRS, *China Report: Agriculture* 258:4-5 (24 May 1983).
28. JPRS, *China Report: Agriculture* 279:14 (17 November 1983).
29. Station Commentary of Wuhan Hubei Provincial Service, "Legal Economic Interests of the Two Households and Economic Bodies Must Be Protected," JPRS, *China Report: Agriculture* 251:106 (2 March 1983). See also "Ministry of Public Security Issues Circular on Protecting Households Who Acquire Wealth Through Labor," JPRS, *China Report: Agriculture* 253:34 (6 April 1983).
30. Li Erzhong, "Dangqian Hebei nongcun shengchan zerenzhi qingkuang di diaocha" (An investigation of the circumstances of Hebei's agricultural responsibility systems at present), *Nongye jingji wenti* 6:12 (1981).
31. "Beijingshi jingji xuehui baochan dao hu zhuanti taolunhui fayan gaoyao" (Notes on comments at the Beijing Municipality Economics Association special discussion meeting on contracting production to the household), *Nongye jingji wenti* 5:61 (1981).
32. Lu Xueyi, "Why Does One Say That *Baochan dao hu* Remains True to Socialism?" JPRS, *China Report: Agriculture* 171:38 (2 November 1981).
33. Anhui and Sichuan provinces, as the provinces first implementing household contracting systems, probably received far greater than proportional shares of input allocations in the past few years.
34. Song Lusheng and Guo Shaoyu, "Shuili jingying guanli shang di yixiang gaige" (A reform in water conservancy management), *Nongye jingji wenti* 9:42-44 (1981).
35. Ma Renping, "Nongcun shixing zerenzhi hou chuxian di xin wenti" (New problems emerging after implementation of responsibility systems in the countryside), *JJGL* 8:4-6 (1981); Zhou Qiren and Wang Xiaoqiang, "On the Changes in the Structure of Economic Functions in the Countryside of Chuxian Prefecture [Anhui]," JPRS, *China Report: Agriculture* 181:30 (30 December 1981).

36. JPRS, *China Report: Agriculture* 236:47 (8 November 1982).
37. See *China Daily*, 14 January 1983, p. 4, "Opinion" column, for information on the development of new "cooperatives" in Zhejiang and Guangdong.
38. Kang Jiusheng, "Zhuanyehu, zhongdianhu zai nongcun jingji zhong di zuoyong ji qi fazhan qushi" (The function of specialized and key-point households in the rural economy, and their developmental tendencies), *Zhongguo jingji wenti* 3:61 (1983).
39. For example, Tianjin Radio Station commentary, "New Economic Combination Should Be Treated Actively and Properly," JPRS, *China Report: Agriculture* 249:154 (16 February 1983).
40. "Draft of the Revised Constitution of the People's Republic of China," *Beijing Review*, 10 May 1982, p. 43; Hu Sheng, "On the Revision of the Constitution," *Beijing Review*, 3 May 1982, pp. 17-18.
41. Mu Qing, Guo Chaoren, and Chen Fuwei, "Zhongguo nongcun di yijue" (A corner of China's countryside), *Hongqi* 4:22-28 (1982); Liu Zheng and Chen Wuyuan, "Nongcun guanli tizhi gaige di chubu changshi" (Initial attempt at the reform of management systems), *JGGL* 4:37-41 (April 1981).
42. Ibid.
43. Ibid., p. 38.
44. Interview with GQW, Hong Kong, June 1983; *People's Daily*, 30 March 1984, p. 1.
45. JPRS, *China Report: Agriculture* 258:29 (24 May 1983), 253:130 (6 April 1983), and 254:100-101 (18 April 1983).
46. *Beijing Review*, 14 September 1981, p. 6.
47. Zhonggong Hunan shengwei zhengyan shi, "Shixing nongshang lianying, gaohuo nongfu chanpin liutong" (Conduct linked management of agriculture and commerce, enliven the circulation of agricultural and subsidiary commodities), *JGGL* 12:66-68 (1982).
48. "CCP Document No. 1," K3-5.
49. On Du's role in policy-making, see Michel Oksenberg, "Economic Policy-Making in China: Summer 1981," *China Quarterly* 90:177 (June 1982).
50. Du Runsheng, "Nongye shengchan zerenzhi yu nongcun jingji tizhi gaige" (The agricultural production responsibility system and reform of the rural economic structure), *Hongqi* 19:22-34 (1 October 1981).
51. Du Runsheng, "Lianchan chengbaozhi he nongcun hezuo jingji di xin fazhan" (The contractual system linked to output and the new development of the rural cooperative economy), *People's Daily*, 7 March 1983, p. 5.
52. Major elements of this vision are contained in "CCP Document No. 1," K1-10. For a synthetic statement of the vision, see Lin Zili, "On the Distinctively Chinese Path of Socialist Agricultural Development," *Social Sciences in China* 3:111-145 (September 1983).

53. JPRS, *China Report: Agriculture* 270:1-2 (2 September 1983); Kathleen Hartford, *The New Stage in Chinese Agricultural Policy* (Columbia East Asian Institute Occasional Papers, forthcoming).
54. I am grateful to Vivienne Shue for the remark that crystallized my thinking on this point.
55. Nicholas R. Lardy, *Agriculture in China's Modern Economic Development* (Cambridge, Cambridge University Press, 1983), p. 208.
56. Ibid., pp. 213-214.
57. Yan Zanyao, "Nongcun xian fuqilai de shi naxie ren?" (Who are the people who have become prosperous first in the countryside?), *People's Daily*, 5 January 1984, p. 2.
58. Wan Li, "Developing Rural Commodity Production," *Beijing Review*, 27 February 1984, p. 19.
59. Interviews with NNU and GQW, Hong Kong, June 1983.

Chapter 2. The Restoration of the Peasant Household as Farm Production Unit in China: Some Incentive Theoretic Analysis, by Louis Putterman

Much of this paper is based on briefings organized by the Ministry of Agriculture, People's Republic of China, during a study tour supported by the Committee on Scholarly Communication with the People's Republic of China, National Academy of Sciences. I would like to thank both the Committee, and the Ministry and other Chinese institutions that contributed to the success of that visit. I also wish to thank the American Council of Learned Societies and the Alfred P. Sloan Foundation for support of research time. In addition, I wish to acknowledge the helpful discussions and suggestions of Marc Blecher, Keith Griffin, Cyril Lin, Barry Naughton, Peter Nolan, Mark Selden, Tom Wiens, and the editors.

1. For an early example of this genre, see Henry Kamm, "Chinese Farms and Factories Thrive on Incentive," *New York Times*, 14 April 1981, p. A9.
2. See Shigeru Ishikawa, "China's Economic Growth since 1949—An Assessment," *China Quarterly* 94:242-281 (1983).
3. See Peter Nolan, "De-collectivisation of Agriculture in China, 1979-82: A Long-Term Perspective," *Cambridge Journal of Economics* 7.381-403 (1983).
4. Karl Marx, *Capital, A Critique of Political Economy*, Vol. 1. (New York, International Publishers, 1967), p. 326.
5. The excessive incentive phenomenon occurs, assuming accurate monitoring of labor, because work points reflect average rather than marginal product. The finding of optimality of incentives under democratic choice is derived in Louis Putterman, "On Optimality in Collective

Institutional Choice," *Journal of Comparative Economics* 5:392–402 (1981). Other relevant theoretical literature includes Dennis L. Chinn, "Team Cohesion and Collective Labor Supply in Chinese Agriculture," *Journal of Comparative Economics* 3:375–394 (1979), and Amartya K. Sen, "Labour Allocation in a Cooperative Enterprise," *Review of Economic Studies* 33:361–371 (1966). The question of the optimal choice of distribution system under collective team production becomes irrelevant when inquiring into the efficiency of the new household contracting systems in which there is no direct team component.

6. There is also the problem of setting ratios of equivalence among *different* tasks, which admits of no absolute solution, but which might be adequately solved in practice by means of social conventions or a collective decision and ratification process.
7. *Rationality* here does not mean "logic" or "logicalness" but, rather, "cognitive capacities."
8. See A. R. Khan and Eddy Lee, *Agrarian Policies and Institutions in China After Mao* (I.L.O. [Asian Employment Programme], Bangkok, 1983; also forthcoming in Keith Griffin, ed., 1984); Nolan; Louis Putterman, "Extrinsic versus Intrinsic Problems of Agricultural Cooperation: Anti-Incentivism in Tanzania and China," *Journal of Development Studies*, 1985.
9. See William L. Parish and Martin King Whyte, *Village and Family in Contemporary China* (Chicago, University of Chicago Press, 1978), p. 116. Basic work points were less discriminating with respect to work completed, although they were probably discriminatory in practice in the sense that, whereas women, e.g., would almost automatically receive fewer points than men under basic work points, some might have been able to complete equal tasks and earn equal points under the task-based system of accounting.
10. It is beyond the scope of the author's current research to offer a definite assessment of the aggregate data series that are relevant to this point, however. For a recent interpretation of evidence on output and consumption, see Nicholas R. Lardy, *Agriculture in China's Modern Economic Development*.
11. Inclusion of this point is not meant to imply that state planning of production is a necessary feature of a socialist agriculture. If one is willing to entertain as socialist the notion of a commonwealth of rural cooperatives embedded in a system of uncontrolled agricultural produce markets, then an institutional structure otherwise resembling that being put in place in China might still qualify, because of collective land ownership.
12. Whereas initially contracts might cover a single crop cycle only, periods of 3 years, 5 years, or longer appeared to be common by late 1983, and, in Rural Work Document no. 1 of 1984, it was announced that the contract period was to be extended to 15 years.

13. The team appears generally to collect the tax from the household and pass it on to the state, but direct payment by the household is also reported.
14. Since obligation to make compulsory deliveries to the state at relatively low prices comes along with contracted land, the implicit tax element in the (below) quota price should be included in the effective rent paid for land use. Its magnitude may have held approximately constant because increases in procurement prices have been counteracted by rising consumer-goods prices facing peasants. As with uncompensated tax in kind, procurement quotas are not rising in general, and certainly not in step with the rising level of output. The government attempts to buy additional grain, when required, at the much higher above-quota and negotiated prices.
15. Interesting exceptions exist in highly industrialized rural areas, where it is reported that households are paid a subsidy by the team to grow grain. In such cases, however, the excess of the tax over the value of marginal product of land might be reversed if units were not *constrained* to grow grain. The subsidies have the same function as did the portions of rural industrial earnings that were included in the values of work points under the old unified distribution system. (In this connection, it may also be worth noting that the loss of the more automatic mechanism for redistributing income between industrial and agricultural workers, which was embodied in the work-point system, is, from a certain viewpoint, a cost of the reforms.)
16. The labor unit for contract land entitlement follows the old basic work-point weightings in some teams, has more moderate dispersion in others, and is equal for male and female workers in still others, according to Keith Griffin (unpublished notes from 1983 study visit to China).
17. On the other hand, a large number of 10,000-yuan households also appear to be former team cadres who have been allocated, or have allocated to themselves, unusually large portions of team resources, on grounds of greater capability to manage these. This phenomenon could undermine the efficacy of rich households as models, since peasants may conclude that new wealth is merely a reward for past political services (and even the implied political road to wealth may be perceived as only a one-time affair to be closed in the future).

Chapter 3. Rural Marketing and Exchange in the Wake of Recent Reforms, by Terry Sicular

1. *Zhongguo tongji nianjian,* 1983 (China statistical yearbook, State Statistical Bureau, Beijing, 1983), p. 159.
2. Ibid., p. 158.
3. Ibid., p. 499. These income figures are in current prices. Real incomes

have probably risen at a slower rate because of inflation in consumer prices over this period. Furthermore, these figures are calculated from rural household survey data. If the growth rate of per capita income distributed by collectives is calculated using regular State Statistical Bureau data, the average annual growth rate from 1978 to 1981 is 11%. See *Zhongguo tongji nianjian,* 1981, p. 198, and *Zhongguo nongye nianjian,* 1980, ed. He Kang (Beijing, 1981), p. 41.

4. For a discussion of the role of rural commerce in economic development, see John C. H. Fei and Gustav Ranis, "Economic Development in Historical Perspective," *American Economic Review* 59.2:386-400 (May 1969).

5. Su Xing, "The Question of Prices of Agricultural Products," *Shehui kexue zhanxian* (Social science front) 2:101-108 (1979), JRPS, *China Report: Agriculture* 58:12-21 (1 November 1979).

6. The particular levels of government administration involved in this planning procedure have varied over time, but the general nature of the procedure has remained unchanged.

7. For a discussion of shifts in the distribution of grain procurement and sales quotas among regions, see Nicholas Lardy, *Agriculture in China's Development,* and Lardy, "Comparative Advantage, Internal Trade and the Distribution of Income in Chinese Agriculture," unpublished manuscript (New Haven, 1982).

8. *Zhongguo caimao bao* (China finance and trade journal), 28 July 1981, p. 2, and 30 July 1981, p. 1.

9. *People's Daily,* 6 October 1979, 6 September 1980, and 30 August 1980; and *Beijing Review,* 15 September 1980, p. 7. In some areas, however, for example, Yunnan and Guizhou, implementation of the promised base-quota and tax reductions was not carried out according to state plan, so that the actual reductions may initially have been less than reported. See Ministry of Agriculture, Bureau of Commune Management, "1977 Zhi 1979 Quanguo Qiongsheng Qingkuang" *Nongye jingji congkan* (A collection on agricultural economics), (1981), reprinted in *Xhinhua yueba* (New China monthly) 2:117-121 (January 1981).

10. Data on recent grain base-quota and tax reductions appear in a number of sources, and the figures given are not completely consistent. For 1979, a figure of 5.5 billion catties (2.75 tons) is given in *People's Daily,* 30 August 1980, p. 1. *People's Daily,* 3 July 1980, p. 1, cites a reduction in the grain tax of 4.735 billion catties (2.37 million tons) for 1979. Grain Ministry Research Bureau, *Tantan nongcun liangyou gouxiao zhengce* (A discussion of rural grain and oil procurement policies; Beijing, 1982), p. 11, gives a figure of 5 billion catties (2.5 million tons) for the 1979 reduction in the base quota and tax. I use the first figure, assuming that the second figure does not include reductions in the base quota, only in the grain tax. Thus, the 0.76-billion-catty difference between the first and second figures should be the additional base-quota reduction. I assume further that the

third figure of 5 billion catties is rounded, and thus consistent with the more precise 5.5-billion-catty figure in the first source. The only figure available for 1980 grain base-quota and tax reductions, and the one I use in this paper, is 1.156 billion catties (0.58 million tons). This appears in the third source above. For 1981, the third source gives a figure of 5 billion catties (2.5 million tons), and *JJNJ*, 1982, p. V-265, gives a figure of 3.55 million tons. I use the latter figure, since it is more precise, and appears in a more recent, authoritative source. Note that these figures are reductions for the grain production and procurement year (April 1 through March 31).

11. *People's Daily*, 27 May 1982, reports that the grain base quota and tax constituted about 40% of total grain procurement in 1981. Given the 1981 state procurement and tax of about 55 million tons trade grain (see Table 2), this implies a 1981 base quota and tax of about 22 million tons. By adding the 6.9 million tons reduction in 1979, 1980, and 1981 to this figure, I estimate a 1978 base quota and tax of about 29 million tons trade grain.
12. Reported by Hubei sources during interviews.
13. FBIS 8 January 1979, L3.
14. Ibid.
15. See note 11 above.
16. Negotiated grain procurement figures are given in Grain Ministry Research Bureau, *Tantan nongcun liangyou gouxiao zhengce*, p. 13.
17. Per capita rural grain consumption in trade grain equivalents was 204.5 kilograms in 1957 and 192.5 kilograms in 1978. See Lardy, *Agriculture in China's Development*, p. 158.
18. Author interview.
19. Author interview.
20. Author interview.
21. Lardy, *Agriculture in China's Development*, p. 158.
22. The impact of prices on production and procurement levels is difficult to predict because price increases do not have an unambiguously positive effect on production and procurement. Higher agricultural procurement prices increase the profitability of agricultural production, thus encouraging production. Price increases, however, also raise rural income. At higher incomes rural households may wish to reduce their labor time and enjoy more leisure, which would have a negative impact on production levels. Moreover, at higher incomes households may wish to retain more of their output for home consumption, which would have a negative impact on procurement levels. If these negative effects outweigh the positive production incentive, then the total output and marketed supply curves will be backward bending.
23. *Zhongguo tongji nianjian*, 1981, p. 405.
24. *Zhongguo nongye nianjian*, 1980, p. 382, and author interviews.
25. Although state quota and above-quota prices appear to have little relation to recent trends in agricultural production and marketing, it is possible that other types of prices may have had some effect. In

particular, state-negotiated procurement prices and market prices could influence production and marketing. In a recent interview, a Chinese economist mentioned that the negotiated procurement price for vegetable oils was set too high, leading to overproduction and high marketing rates of oil crops. Rural market prices could also explain some of the trends in production and marketing. Unfortunately, no negotiated price data and insufficient rural market price data are available with which to carry out further analysis on this topic.

26. *JJNJ*, 1982, ed. Xue Muqiao (Beijing, 1983), p. V-276.
27. Alexander Eckstein, *China's Economic Revolution* (London, Cambridge University Press, 1977), p. 117. Free market sales are the difference between columns 3 and 2 in Eckstein's Table 4-1. For the sake of consistency, I use the procurement data in Table 4-1 to calculate free market sales as a percentage of state procurement, rather than using procurement data presented earlier in this paper.
28. FBIS, 25 May 1983, K5, and *Beijing Review*, 1 August 1983.
29. *People's Daily,* 10 March 1980, and 9 October 1980.
30. *Zhongguo nongye nianjian,* 1980, p. 326, gives a figure of 58.68 billion yuan for the total value of state agricultural procurement in 1979. In 1980, state agricultural procurement rose to 67.7 billion yuan. *JJNJ,* 1981, p. IV-121.
31. FBIS 8 May 1981, K7, and *Beijing Review,* 18 May 1981, pp. 7-8.
32. *Zhongguo caimao bao,* 11 August 1981, pp. 1-2, and FBIS 9 June 1982, K8.
33. *Zhongguo caimao bao,* 6 January 1981, p. 1.
34. *People's Daily,* 10 January 1981, p. 1; *Zhongguo caimao bao,* 11 August 1981, pp. 1-2; and FBIS, 9 June 1981, K8.
35. FBIS 30 March 1982, K5.
36. *People's Daily,* 17 May 1984, p. 2.
37. *People's Daily,* 9 April 1981 (also in *Zhongguo caimao bao,* 11 April 1981, p. 1); FBIS 31 August 1982, K15; and *Beijing Review,* 1 August 1983, p. 4.
38. FBIS 25 May 1983, K5.

Chapter 4. Getting Rich Through Diligence: Peasant Income After the Reforms, by S. Lee Travers

1. The use of modern inputs purchased through state trading channels, which monopolize such inputs, increased at an average annual real rate of 11% from 1957 to 1977, and the total cost of production increased at 6.7% annually over this period, exclusive of labor. With 1957=100, the overall agricultural purchase price index was 137.8 in 1961 and 143.1 in 1977. See Nongyebu zhengce yanjiushi, ed., *Zhongguo nongye jingji gaiyao* (Outline of China's agricultural economy; Beijing, 1982), pp. 98, 191; Guojia Tongjiju, ed., *Zhongguo tongji nianjian 1981* (Statistical yearbook of China, 1981; Beijing, 1982), pp. 195, 403.

Peasant population grew at 2.0% annually; Zhongguo nongye nianjian bianji weiyuanhui, ed., *Zhongguo nongye nianjian 1980* (Statistical Yearbook of Chinese Agriculture, 1980; Beijing, 1981), pp. 5, 319.
2. A State Statistical Bureau (SSB) report gives 1956 peasant per capita income as 73 yuan; *Guangming ribao*, 7 February 1981, p. 1. Using the implied inflation adjustment given by Chinese economist Zheng Linzhuang yields the 102.8 yuan figure; Zheng Linzhuang, "Nongye xiandaihua yu nongye shengchan xiaolu" (Agricultural modernization and agricultural production efficiency), *Zhongguo shehui kexue* (Chinese social sciences) 2:3-16 (1981).
3. All calculations were done on deflated data, derived from the SSB; Guojia tongjiju, ed., "Zhongguo jingji tongji ziliao xuanbian" (Selected material on China's economic statistics) in *JJNJ* (Almanac of China's Economy, 1982), ed. Zhongguo jingji nianjian bianji weiyuanhui (Beijing, 1983), pp. VIII-23, 24, 26. The definition of *nongcun* changed in 1963; Guojia Tongjiju, ed., *Zhongguo tongji nianjian 1981*, p. 495. We lack data that would establish the implications of that change for comparisons of income and consumption before and after 1963. The *nongcun* category covers more than just peasants; rural consumption figures given in the text overestimate peasant consumption (see note 60 below for a more thorough discussion of the *nongcun* category). Purchases of consumer goods are obviously not a measure of total income, but rather that part of income used to purchase commodities. It is, however, a useful measure of relative buying power. It measures total urban income more closely than rural, as nearly all urban income is realized in cash, while much rural income is consumed in kind.
4. See Hua Guofeng, "Unite and Strive to Build a Modern, Powerful Socialist Country!" in *Documents of the First Session of the Fifth National People's Congress of the People's Republic of China* (Beijing, 1978), pp. 42-50.
5. *JJNJ*, p. IV-23; Guojia tongjiju, ed., *Zhongguo tongji nianjian 1981*, p. VIII-26.
6. For details of methodology and data sources, see Travers, "Post-1978 Rural Economic Policy" (forthcoming), Appendix B.
7. Liangshibu Yanjiushi, ed., *Tantan nongcun liangyou gouxiao zhengce* (A discussion of rural grain and oil purchase and sales policy), p. 11, and Liang Yan, "Zhongguode liangshi fenpei he guanli" (Chinese grain distribution and management) in *JJNJ 1982*, pp. V-262-265.
8. Author interview.
9. Liangshibu Yanjiushi, p. 13.
10. Wu Zhenkun, "Guanyu nongye jianchi jihua jingji weizhude jige wenti" (On several problems of agriculture following the planned economy), *People's Daily*, 27 May 1982, p. 5.
11. Travers, "Post-1978 Rural Economic Policy" (forthcoming), Table B5.
12. Ibid.

13. Zhao Ziyang, "Guanyu dangqian jingji gongzuo de jige wenti" (On several problems in current economic work) *People's Daily*, 30 March 1982, p. 3.
14. Zhongguo Nongye Nianjian Bianji Weiyuanhui, ed., *Zhongguo nongye nianjian 1980* (Beijing, 1981), p. 366; Table 3, p. 19; Travers, "Post-1978 Rural Economic Policy" (forthcoming), Table A3.
15. Guowuyuan, "Guanyu fazhan sheduiqiye ruogan wentide guiding (shixing caoan)" (Regulations on several problems in the development of commune and brigade enterprise [trial draft]), in *JJNJ 1981*, p. II-98.
16. "Guowuyuan guanyu tiaozheng nongcun sheduiqiye gongshangshuishou fudande ruogan guiding" (Several State Council regulations on the adjustment of the incidence of the industrial and commercial tax collection in rural commune and brigade enterprise), in *JJNJ 1982*, p. III-55.
17. "Guowuyuan guanyu sheduiqiye guanche guomin jingji tiaozheng fangzhende ruogan guiding" (Several State Council regulations on commune and brigade enterprise implementation of policies on the readjustment of the national economy), in *JJNJ 1982*, pp. III-13-15.
18. Guojia Tongjiju, ed., *Zhongguo tongji nianjian 1981*, p. 192.
19. Ibid., p. 195.
20. Liangshibu Yanjiushi, p. 12.
21. Audrey Donnithorne, *China's Economic System* (London, Allen and Unwin, 1967), p. 361, and Liangshibu Yanjiushi, p. 48.
22. "Decisions of the Central Committee of the Communist Party of China on some questions conerning the acceleration of agricultural development (draft)," *Issues and Studies* 15.7:112 (July 1979). The revised version of this document can be found as "Zhonggong zhongyang guanyu jiakuai nongye fazhan ruogan wenti de jueding," in *JJNJ 1981*, pp. II-100-107.
23. Shen Jingnong and Chen Baosen, "1980 nian de Zhongguo caizheng" (Chinese public finance in 1980), in *JJNJ 1981*, p. IV-151.
24. Guowuyuan, "Guowuyuan pizhan caizhengbu guanyu zhixing nongye shui qizhengdian banfa de qingkuang baogao" (The State Council approves and transmits a Ministry of Finance report on the situation in implementing the method of an agricultural tax basic exemption), *Zhonghua Renmin Gongheguo Guowuyuan gongbao*, 1980 (Public announcements of the State Council of the People's Republic of China) 13:3 (26 August 1980).
25. Nongyebu zhengce yanjiushi, ed., *Zhongguo nongye jingji gaiyao*, p. 198.
26. "Decisions of the Central Committee of the Communist Party of China on some questions concerning the acceleration of agricultural development (draft)," p. 112.
27. Zhongguo nongye nianjian bianji weiyuanhui, ed., *Zhongguo nongye nianjian 1980* (Beijing, 1981), p. 262.

28. Guojia tongjiju, ed., "Zhongguo jingji tongji ziliao xuanbian," p. VIII-26, and Guojia tongjiju, ed., *Zhongguo tongji nianjian 1981*, p. 329.
29. Provincial grain exports were 7 to 7.5 mmt per year in the 1953-1956 period, falling to 4.7 mmt in 1965 and 2.05 mmt in 1978 (Nicholas Lardy, *Agriculture in China's Development*, Table 2.2). As a percentage of total output the fall was even greater. In addition, grain sales to peasants fell absolutely between the 1953-1958 period and the late 1970s (ibid., Table 2.1; author interview).
30. "Decisions of the Central Committee . . . on the acceleration of agricultural development (draft)," p. 110.
31. Zhang Siqian, "1980 niande Zhongguo nongye" (China's agriculture in 1980), in *JJNJ*, pp. IV-11-18.
32. Acreage sown to grain crops fell 6.6 million hectares from 1978 through 1981, while grain output increased 20.3 mmt over those crop years; *China Daily*, 29 December 1981, p. 3; Guojia tongjiju, ed., "Zhongguo jingji tongji ziliao xuanbian," p. VIII-4.
33. Liang Yan, p. V-265.
34. Liangshibu yanjiushi, pp. 12, 52; Lardy, Chapter 2.
35. Xu Daohe, "Nongye ye yao yi jihua jingji weizhu shichang tiaojie weibu" (Agriculture must also take the planning economy as primary and market adjustment as secondary), *Lilun yu shijian* 3:8-10 (1982).
36. "Decisions of the Central Committee . . . on the acceleration of agricultural development (draft)," p. 108.
37. "Regulations on the work in the rural people's communes (Draft for trial use)," *Issues and Studies* 15.9:105 (September 1979).
38. Those problems were not new; they had arisen as early as the mutual-aid-team movement in the early 1950s. See Vivienne Shue, *Peasant China in Transition*, (Berkeley, University of California Press, 1980), pp. 163-176.
39. Zhou Yueli, "Anhui sheng de nongye shengchan zerenzhi" (The agricultural production responsibility system in Anhui province), in *JJNJ 1981*, pp. III-62-64.
40. *People's Daily*, 29 August 1981, p. 2.
41. Zhang Siqian, pp. IV-13-14.
42. Zhang Siqian, p. IV-13; Zhou Cheng, "Lun baochan dao hu" (A discussion of giving quotas to households), *Jingji lilun yu jingji guanli* 2:40 (1981).
43. Ai Yunhang, "1981 niande Zhongguo nongye" (China's agriculture in 1981), in *JJNJ 1982*, pp. V-11-12.
44. "Decisions of the Central Committee . . . on the acceleration of agricultural development (draft)," pp. 108, 111.
45. The free markets are also used as a sales outlet by the collective economy, though that apparently was not intended in 1978, when it was stated, "Products of the (collective) side-occupations should be centrally marketed." "Regulations on the work in the rural people's communes (Draft for trial use)" 15.8:105 (August 1979).

46. Ibid., 15.9:110.
47. No information on relative input use was given. The official conclusion seems to be that better management alone accounted for the observed yield differences. Differences in the intensity of physical inputs (fertilizer, for example) is another likely explanation of the results. *People's Daily,* 17 June 1981, p. 2.
48. Li Bingkun, "Zhongguo nongcun sheyuan jiating fuye" (Household sidelines of China's peasants) in *JJNJ 1982,* p. V-20; "Dangzhongyang guowuyuan fachu tongzhi yaoqiu jiji fazhan duo zhong jingying" (An instruction of the Party Central Committee and State Council calling for active development of all-around management), *People's Daily,* 4 June 1981, p. 1.
49. Li Bingkun, p. V-20; Liu Zhongyi and Liu Yaochuan, "Nongye Jiegou" (The structure of agriculture), in Ma Hong and Sun Xiangqing, eds., *Zhongguo jingji jiegou wenti yanjiu,* 2 vols., (Beijing, 1981), p. 145; Ma and Sun, eds., p. 145; Ma Hong, ed., *Xiandai Zhongguo jingji shidian* (Reference work on the modern Chinese economy; Beijing, 1982), p. 120.
50. Zhang Siqian, p. IV-14.
51. Wang Yiping, "Kaifang nongcun jishi maoyi" (Liberate rural free-market trade), in Zhongguo nongye nianjian bianji weiyuanhui, ed., *Zhongguo nongye nianjian 1980,* p. 158; Shangyebu shangye yanjiusuo, "1980 niande Zhongguo shangye" (China's commerce in 1980) in *JJNJ 1982,* p. IV-122.
52. Liang Yan, p. V-264; Guojia tongjiju, ed., "Zhongguo jingji tongji ziliao xuanbian," p. 341.
53. Travers, "Post-1978 Rural Economic Policy," Table A2.
54. Guojia tongjiju, ed., *Zhongguo tongji nianjian 1981,* p. 409.
55. Ibid., ed., p. 198; Nongyebu zhengce yanjiushi, ed., *Zhongguo nongye jingji gaiyao* (Outline of China's agricultural economy), p. 202.
56. Guojia Tongjiju, ed., *Zhongguo tongji nianjian 1981,* p. 199; Liang Yan, "Lun wo guo nongye de shangpinhua" (On the commercialization of Chinese agriculture), *Caijing yanjiu* 3:33 (1981).
57. See Travers, "Post-1978 Rural Economic Policy," Table B6.
58. Ibid., Table B6.
59. Ibid., Table B5.
60. With 1978 as the base year, total consumer goods purchases are adjusted using a common deflator for both urban and rural groups; Travers, "Post-1978 Rural Economic Policy," Table A4. The main reason for the lower rural inflation figures reported in the SSB annual reports is the near zero inflation in agricultural producer goods, which make up nearly a quarter of rural retail sales.

Recent data show the degree to which the *nongcun,* or rural, category of retail sales overestimates sales to peasants. Since 1964, *chengzhen,* or urban, is defined as an area with more than 3,000 permanent residents, 70% or more of whom are not dependent on agricultural labor for their income. All other places are *nongcun*, as is the agricultural population in urban areas (Guojia tongjiju, ed., *Zhongguo tongji nianjian 1981*, p. 495). Approximately 28 million people in rural areas receive commercial grain rations, hence are not peasants; Liang Yen, p. V, 263; Guojia tongjiju, ed., *Zhongguo Tongji Nianjian 1981*, p. 3. Their local purchases will be recorded as *nongcun* sales. Peasants also purchase goods in cities, of course, and the net effect of these two factors on total peasant purchases from the commercial system is unknown.

61. Table 14 does not compare total consumption, as it does not include services or consumption-in-kind. If adjustments for these factors, and price differences, were made, total consumption across sectors would show less difference. Peasant purchases of retail goods will increase at a much faster rate than total peasant income because the marginal rate of retail purchases from income is far above the average rate.
62. This includes both urban and rural populations; Gui Shiyong, "Cong yiqie wei renmin de sixiang chufa, tongchou anpai shengchan jianshe he renmin shenghuo" (Starting from the ideology of serving the people, plan construction for production and people's life), in *Zhongguo jingji nianjian 1982*, p. IV-80.
63. Or that the production system is producing goods desired in rural areas but not urban. In either case, the command system is effecting the redistribution; it is not simply a matter of increased relative income in rural areas.
64. National income figures for agriculture, deflated by the agricultural purchase price index on a 1978 base, were used to derive the growth rate; Guojia tongjiju, ed., *Zhongguo Tongji Niajian 1981*, pp. 20, 403. Gross output, valued in 1970 yuan, has averaged a 5.6% rate of growth over the same period; Guojia tongjiju, ed., "Zhongguo jingji tongji ziliao xuanbian," p. VIII-9.
65. Xu Daohe, p. 9.
66. *People's Daily*, 20 June 1981, p. 1.

Chapter 5. Politics, Welfare, and Change: The Single-Child Family in China, by Joyce K. Kallgren

I wish to acknowledge the great assistance given me in research for this article by Mr. C. P. Chen and Mrs. Annie Chang of the Center for Chinese Studies and Mr. Lau of the Universities Service Centre, Hong Kong. I received financial support from the Committee of Research (Davis campus) and the Center for Chinese Studies (Berkeley campus), of the University of California.

1. *People's Daily*, 2 February 1982.
2. *Zhonghua renmin gongheguo faling huibian* (Collection of laws of the People's Republic of China, 1950-1963), title varies, 1955, pp. 708-710.

3. H. Yuan Tien has written a short, well-researched and documented study entitled, "China: Demographic Billionaire," *Population Bulletin* 38:1–43 (April 1983). Major demography journals have published studies discussing available data, such as Leo Goodstadt's "China's One Child Family: Policy and Public Response," *Population and Development Review* 8.1:37-58, 232-234 (March 1982). Janet Salaff addresses important issues in "The Right to Reproduce: The People's Republic of China," a paper presented for the Conference on Reproduction Rights and Responsibilities at the University of Missouri-Columbia in June 1983. Tyrene White uses field data acquired in China to outline a number of issues related to the policy—"Implementing the One-Child-per-Couple: Population Program in Rural China: National Goals and Implementation," a paper prepared for the Workshop on Policy Implementation in Post-Mao China, Columbus, June 1983. Elisabeth Croll, in her 1983 volume, *Chinese Women Since Mao* (M. E. Sharpe), has a short but excellent summary of issues in this problem, pp. 88-104. A working paper by Professor Wong Siu-lun, "China's Present Population Policy: Some Likely Consequences," was presented to the Contemporary China Studies Seminar Programme of the Centre of Asian Studies of the University of Hong Kong; the article has been published in the *China Quarterly*, June 1984, pp. 220-240, with the revised title "Consequences of China's New Population Policy."
4. *Dongbei laobao jingyan jieshao* (Introduction to the Northeast experience in labor insurance), compiled by the Shanghai Trade Union Federation, Labor Protection Bureau, 1950, pp. 1-13.
5. "Report of a Visit to the People's Republic of China, 1976," prepared for the House of Representatives, Committee on International Relations, 84th Congress. Second Session.
6. See *Dongbei laobao* for benefit scales and advantages that accrued to trade-union members.
7. "Report of a Visit," pp. 36-38.
8. "China's Population Policy and Population Data, Revised and Updated 1981," prepared for the Committee on Foreign Affairs, U.S. House of Representatives, by the Congressional Reference Service Library of Congress, p. 1.
9. P. C. Chen, *Population and Health Policy In the People's Republic of China* (Washington D.C., Smithsonian Institution, Occasional Monograph Series no. 9).
10. Liu Zheng, "China's Population Program," *Renkou Yanjiu* (Population Studies) 1982, JPRS 295:28 (29 April 1982).
11. William L. Parish and Martin K. Whyte, *Village and Family in Contemporary China* (Chicago, University of Chicago Press, 1978), pp. 131-154, especially pp. 324-325.
12. Zhao Ziyang, "Work Report," *Beijing Review* No. 27 (1983), pp. II-XXXIII, and No. 24 (1984), pp. I-XVI.

13. *Population and Birth Planning in the People's Republic of China,* Population Report Series J no. 25, January–February 1982, pp. J-582.
14. *People's Daily,* 7 March 1980.
15. *People's Daily,* 21 January 1982. Since adoption of the Single-Child-Family policy, the policy has been supported by pictures of harried mothers with willful and possibly poorly cared-for children, but it is my view that these developments are only ancillary to the arguments of state need.
16. Geoffery McNicoll, "Institutional Determinants of Fertility Change," in *Determinants of Fertility Trends: Theories Re-examined,* ed. C. Hahman and R. McKenzie (Liège, International Union for the Scientific Study of Population Liège Ordina Ed. 1982), p. 443.
17. Judith Bannister, "Strengths and Weaknesses of China's Population Data," a paper presented to the China Population Analysis Conference at the East West Center, May 1980, to be published in a volume by Stanford University Press (forthcoming). This is a first-rate piece of scholarship.
18. *Guangming ribao,* 12 January 1980, JPRS no. 89 (9 June 1980).
19. A. J. Coale, "Population Trends, Population Policy and Population Studies in China," in *Population and Development Review* 7.1:85-97.
20. Liu Zheng, Wu Canping, and Lin Fuda, "Five Recommendations for Controlling Population Growth in China," *Renkou yanjiu,* JPRS 180:5 (8 April 1981).
21. Chen Muhua, "Develop Population Science and Make It Serve the Goal of Authority Controlling Population Growth," *Renkou yanjiu* 3:1-7 (1981).
22. *Xinhua,* Beijing, "Family Planning Month Activities Start," FBIS 13 January 1983, K-13.
23. *People's Daily,* 27 January 1979, 30 June 1979, 24 December 1979.
24. *People's Daily,* 31 March 1981.
25. Author's personal interview with an obstetrician-gynecologist in Changsha who was conducting such research (May 1979).
26. *People's Daily,* 11 February 1980.
27. *Renkou yanjiu* 2:2 (1978).
28. Huang Xuelin, "Economic Construction and Our Country's Population Increase," 1979, JPRS PS&M no. 90:93 (13 June 1980).
29. *Renkou yanjiu* 5:29 (1981).
30. *Renkou yanjiu* 5:29 (1982).
31. *Renkou yanjiu* 5:29 (1981).
32. Liu Haiquan, "Control Population Growth with the Same Drive Used in Grasping Production and Construction," *Guangming ribao,* 13 September 1979, JPRS 39:68-71 (3 December 1979), Huang Zuelin.
33. W. R. Lavely, "China's Rural Population Statistics at the Local Level," *Population Index* 1982, XXXVIII, 665-677.
34. T. White, p. 26.

35. *People's Daily*, 6 March 1979, 5 June 1980.
36. *Main Documents of the Third Session of the Fifth National People's Congress of the People's Republic of China*, Beijing 1980 Appendixes, "The Marriage Law."
37. Zhu Yuncheng, "The Importance and Urgency of Promoting the Policy of One Child for Every Married Couple in Our Country as Viewed from the Point of View of Development," JPRS SEC no. 167 (3 March 1981). Xu Dixin et al., *China's Search for Economic Growth* (Beijing, New World Press, 1982), pp. 1-9.
38. *Population and Birth Planning*, pp. J-6-5.
39. *Nanfang ribao*, 5 January 1983.
40. Nanchang, Jiangxi, "We Should Strictly Adhere to the Policy and Regulations Permitting Women to Bear a Second Child if a Second Child is Justified According to Regulations," FBIS 10 December 1982.
41. *Population and Birth Planning*, p. J-589; *Shaanxi, nongmin bao*, 8 January 1983.
42. *Shaanxi nongmin bao*, 8 January 1983.
43. *Tianjin ribao*, 20 January 1983.
44. T. White, pp. 17-19, refers to the penalties as a tax of 200 yuan for the second child and 300 yuan for the third child.
45. Joan M. Maloney, "Recent Developments in China's Population Planning," *Pacific Affairs* 54.1:108. The most graphic account on this problem is to be found in Stephen Mosher, *Broken Earth: The Rural Chinese* (Free Press, 1983), pp. 224-261. The author also found this impression confirmed in interviews in the summer of 1983 in Hong Kong.
46. *Sichuan ribao* 22 January 1983; *Renkou yanjiu* 2:70 (1981).
47. *Shaanxi ribao*, 21 January 1983. Apparently, there are ongoing efforts to get physicians to certify sterilization fraudulently. Sometimes the efforts are successful; *Zhejiang ribao*, 24 January 1984. Sometimes they are not; *Zhejiang ribao*, 12 October 1983.
48. Victor Nee, "Post Mao Changes in a South China Production Brigade," *Bulletin of the Committee of Concerned Asian Scholars* 13:32-40 (1981).
49. T. White, p. 29; also see the experience in Pu Xian reported in *Renkou yanjiu* 5:29 (1982).
50. C.C. Ching, "The One Child Family in China: The Need for Psychosocial Research," *Studies in Family Planning* 13.3:209 (June-July 1982); also see *Shaanxi nongmin bao*, 8 January 1983.
51. *Guangming ribao*, 7 September 1979.
52. T. White, pp. 29.
53. *Renkou yanjiu* 6:29 (1982).
54. *Renkou yanjiu* 2:47, 61 (1982). This report includes a list of various alternative contract options that have been developed to integrate the responsibility system with birth-planning targets. In some cases, a single target combines both production and birth goals; in other cases parallel contracts are signed but with integrated targets.

55. Huang Xuelin, p. 95.
56. Zhang Huaiyu and Dong Shigui, "New Problems in the Control of Population in the Rural Areas Brought About by Implementation of the Agricultural Responsibility System," *Renkou yanjiu*, 29 January 1982, in JPRS.
57. Gui, pp. 32.
58. *Sichuan ribao*, 19 January 1983; *Haerhbin ribao*, 14 January 1983; *Guangming ribao*, 17 April 1983; *People's Daily*, 31 January 1983.
59. Salaff, p. 35. This matter should be treated with great care, since there is so much contradictory evidence. For example, in Fei Hsiao Tung (Fei Xiaotung) *Chinese Village Close-Up* (New World Press, Beijing, 1983), pp. 261, there is a report of a research team that reads as follows: "Kaixian gong has a tradition of limiting population by means of abortion and infanticide. The average number of children per couple is not high." This observation about a historical problem is not linked to the SCF campaign.
60. *People's Daily*, 31 January 1983; *Guangming ribao*, 20 February 1983.
61. Jurgen Domes, "New Policies in the Communes: Notes on Rural Societal Structures in China," *Journal of Asian Studies* 41.2:253-267 (February 1982).
62. Joyce K. Kallgren, "Daily Life and Social Policy" in *China After Thirty Years*, ed. J.K. Kallgren (Berkeley, Institute of East Asian Studies, University of California, Berkeley, 1979), pp. 95-120.
63. Zheng Liu, "A Preliminary Analysis of a Planned Birth Program in a Chinese Farming Community—a Planned Birth Field Study in Shifang County, Sichuan Province" (unpublished paper, May 1980).
64. *Zhonghua nongmin pao*, 5 May 1983.

Chapter 6. The Implications of Rural Reforms for Grass-Roots Cadres, by Richard J. Latham

1. Feng Jixin, "Something Imperative—A Survey of Production-Responsibility Systems in the Rural Areas," *People's Daily*, 11 July 1981, p. 2.
2. Beijing domestic service, 25 November 1981, FBIS 3 December 1981, O2.
3. Fang Yuansheng, "Rural Areas of Guangdong Enter 'Second Golden Age,'" NCNA (Beijing), 18 August 1982, FBIS 19 August 1982, K5-7.
4. Dong Qiwu, "Poverty Is Changed into Prosperity; the Country Is Prosperous and the People Live in Peace—The Party's Wise Leadership As Seen from Three Investigations Conducted in Rural Areas," *People's Daily*, 27 June 1982, p. 2.
5. "Hebei Provincial CCP Committee Rural Work Department's 'Opinions on Regularly Grasping the Work of Perfecting the Agricultural Production-Responsibility Systems,'" *Hebei ribao*, 12 June 1981, p. 1, FBIS 9 July 1981, R2.

6. "Enthusiastically Assist Poverty-Stricken Communes and Production Brigades and Teams," *People's Daily*, 18 May 1982, p. 1.
7. Jin Wen, "The Agricultural Production Responsibility System and the Socialist Road," *Guangming ribao*, 25 April 1981, p. 4.
8. Bai Dongcai, "Adopt a Correct Attitude Toward the Responsibility System of Assigning Land to Each Household in Exchange for Specific Levies and Fixing Output Quotas Based on the Household," *People's Daily*, 29 June 1982, p. 2.
9. Du Runsheng, "A Historic Change in Rural Work," *People's Daily*, 16 September 1982, p. 4.
10. "Don't Arbitrarily Divide Up the Production Teams," *Nanfang ribao* (Guangzhou), 1 March 1979 FBIS, 2 March 1979, H1; FBIS 10 March 1979, H1-2; FBIS 23 March 1979, P4-5; FBIS 21 March 1979, H1.
11. FBIS 5 December 1980, S1; Jin Wen, p. 4; FBIS 7 May 1981, O2; Feng Jixin, p. 2; FBIS 3 September 1981, K9; Yao Liwen and Xu Xiji, "The Rural Collective Economy Gains New Vitality—Report on a Visit to Rural Guizhou," *People's Daily*, 5 June 1982, p. 2; FBIS 13 August 1982, S6; Zhou Zijiang, "The Spirit of the Third Plenary Session Has Guided Us to Realize the Great Historical Change," *People's Daily*, 3 September 1982, p. 3; He Rongfei, "The Agricultural Production Responsibility System and the Problems of Impoverished Households," *Guangming ribao*, 25 September 1982, p. 3.
12. "What Has 'Happened?'—Written to Dispel the Doubts of Rural Cadres," *Liaoning ribao* (Shenyang), 22 February 1982, p. 1, FBIS 17 March 1982, S1; "Policies Must Be Consistently Stable," *People's Daily*, 3 April 1982, p. 1; FBIS 17 November 1982, P9-10.
13. CCP Central Committee Administrative Office and CCP Central Party School Investigation Group, "An Investigation of and Views about Several Present Rural Systems Whereby Responsibility Is Linked to Output," *People's Daily*, 1 September 1981, p. 2.
14. Ibid.; "The Road They Take Leads to Common Prosperity," *Jiefangjun bao* (Liberation Army news), 22 October 1980, FBIS 23 October 1980, L3-4; "Why Did a Few Comrades Misunderstand and Doubt the Party's Policies at One Time?—Leading Cadres of a Certain Army Unit of the Guangzhou PLA Units Examine Causes for Problems in Implementing the Ideological Line, *People's Daily*, 17 November 1980, p. 4.
15. "What Has 'Happened?'" S1.
16. CCP Central Committee Administrative Office and Central Party School, "Several Questions of Cognition on Perfecting the Agricultural Production Responsibility System," *People's Daily*, 27 April 1982, p. 2.
17. The respondent data in Tables 19 and 20 must be qualified in three ways. First, the respondents were seldom aware of the personal views of commune-level cadres regarding the impact of the rural reforms. Second, none of the respondents resided in China after the CCP Central Committee and State Council announced far-reaching plans to

separate the government administrative and production functions of communes. And, third, only two of the respondents were leading cadres in their production teams. The respondents reported, therefore, their perceptions of cadre reactions or, in several instances, the views of relatives who were cadres. By drawing upon published accounts of cadre reactions, it has been possible to show that those perceptions were not out of line with public, frequently official, observations.

18. NCNA, "Correctly Understand and Treat the Separation of Government Administration from Commune Management," 1 June 1982, FBIS 3 June 1982, K3.
19. Wu Xiang, "The Open Road and the Log Bridge—A Preliminary Discussion on the Origin, Advantages and Disadvantages, Nature, and Future of the Fixing of Farm Output Quotas for Each Household," *People's Daily,* 5 November 1980, p. 2.
20. Feng Jixin, p. 2; NCNA, "Work Study: Reform the System of Rural Economic Management, Strive to Lighten Peasants' Burdens," 25 August 1981, FBIS 26 August 1981, K1-3.
21. Feng Jixin, p. 2.
22. "Strictly Carry Out the Activities of Party Organizations from Top to Bottom, *People's Daily,* 5 May 1981, FBIS 6 May 1981, K8.
23. "Cadres Must Do a Better Job Under the Responsibility System," *Fujian ribao* (Fuzhou), 21 March 1982, p. 1, in JPRS/CRPSMA 298:84 (12 May 1982).
24. "Shanxi Provincial CCP Committee Issues Supplementary Circular on Further Perfecting the Agricultural Production Responsibility System," *Shanxi ribao* (Taiyuan), 29 November 1981, p. 1, FBIS 15 December 1981, R3.
25. Interviews 5GR, 10GR, and 20AR.
26. Interview 7GR.
27. "Hebei Provincial CCP Committee Rural Work Department," R2.
28. FBIS 3 June 1982, K3.
29. A persistent but not always apparent objective of the Third Plenum economic reforms was the reduction of costly, non-production personnel in rural as well as urban areas. From Guangdong to Gansu and Jilin the themes were the same in 1981 and 1982; "Reduce the number of personnel so as to lighten the masses' burdens"; "cut irrational expenses"; and keep remuneration at a "reasonable level." Liu Yuzhai, secretary of the Xinyang prefectural CCP committee in Henan, decried the extreme costliness of supporting large numbers of non-production personnel in his prefecture. He cited data for the Liuhe production brigade to illustrate his point. In 1979, the brigade supported 146 non-production personnel at a cost of 32.95 yuan and 144 jin of grain per brigade member. By 1980, non-production personnel were reduced to 42 and only 21 of them received any "non-production subsidies." The cost of supporting these cadres was reduced from 42,968 yuan in 1979 to 3,264 yuan in 1980. See FBIS 26 August 1981, K1-2.

30. Feng Jixin, p. 2.
31. "State Farms Must Strive to Raise the Management Level," *People's Daily*, 22 April 1982, p. 2.
32. For a discussion of the various kinds of rural production responsibility systems, see Greg O'Leary and Andrew Watson, "Current Trends in China's Agricultural Strategy: A Survey of Communes in Hebei and Shandong," *The Australian Journal of Chinese Affairs* 8:10-16 (1982); Su Xing, "Responsibility Systems and the Development of the Rural Collective Economy," *JJGL* 11:6 (1982); and David Zweig, "Opposition to Change in Rural China: The System of Responsibility and People's Communes," *Asian Survey* 23.7:882-884 (July 1983).
33. A "mixed production responsibility system" refers to the existence of several different responsibility systems within a production brigade and commune. This situation often resulted in difficult and troublesome management problems. For example, some respondents observed that, in their communes or brigades, production teams that had decided to assume responsibility for all production tasks and costs ceased contributing to collective medical services. For a fee, however, individuals could still utilize the facilities that were maintained by other teams in the brigade and commune which remained under some form of collective control.
34. "Properly Solve the Problem of Cadres, Staff Members and Workers Returning Home to Cultivate Farmland for Which They Are Responsible," Hunan provincial service, 24 February 1982, FBIS 25 February 1982, P2.
35. FBIS 7 April 1982, K4.
36. "It Is Necessary to Show More Concern for and Assist Grass-Roots Cadres in Rural Areas," *People's Daily*, 15 July 1981, pp. 1, 4.
37. FBIS 10 July 1981, K1; "Regional CCP Committee Arranges the Next Stage of Work; Demands That Three Things Be Tackled Well at the Forum of Cadres Sent Down to Rural Areas," *Ningxia ribao* (Yinchuan), 8 April 1982, p. 1; FBIS 30 April 1982, T2.
38. "Hebei Provincial CCP Committee Rural Work Departments," R5.
39. Interview 3GR.
40. FBIS 7 April 1982, K12.
41. Feng Jixin, p. 2.
42. FBIS 8 January 1981, T1.
43. Interview 3GR.
44. "Shanxi Provincial CCP Committee Issues Supplementary Circular," R3.
45. "Hebei Provincial CCP Committee Rural Work Departments," R5.
46. FBIS 7 August 1979, P3.
47. FBIS 3 March 1981, L24.
48. Jin Wen, 15 April 1981, p. 4.
49. "Seriously Do a Good Job in Rectifying and Building the Rural Basic

Level Leadership Groups," *Hebei ribao* (Shijiazhuang), 12 November 1981, p. 1, FBIS 9 December 1981, R2.
50. "Seriously Overhaul the Financial Affairs of Communes," *People's Daily*, 12 December 1980, p. 2; "Hebei Provincial CCP Committee Rural Work Department," R5; "It Is Necessary to Show More Concern for and Assist Grass-Roots Cadres in Rural Areas," *People's Daily*, 15 July 1981, p. 1, FBIS 22 July 1982, P6; "Seriously Do a Good Job in Rectifying and Building the Rural Basic Level Leadership Groups," R3.
51. "On Further Strengthening and Perfecting the Agricultural Production Responsibility System," *Shaanxi ribao*, 7 November 1980, FBIS 12 November 1980, T2.

Chapter 7. Rural Collective Violence: The Fruits of Recent Reforms, by Elizabeth J. Perry

Special thanks for a critical reading of this manuscript in draft form are due to John Burns, Jean Oi, James Townsend, and Christine Wong.
1. A more detailed treatment of violence during the first decade of the People's Republic is contained in Elizabeth J. Perry, "Rural Violence in Socialist China," (paper presented at the workshop on "Recent Reforms in China," Harvard University, April 1983).
2. William Parish and Martin K. Whyte, *Family and Village in Contemporary China* (Chicago, University of Chicago Press, 1978), p. 302.
3. Ibid., pp. 116-117, 327-328.
4. Contrast the situation after collectivization with the situation just after land reform when, as Thomas P. Bernstein has shown, the desire to withdraw from local leadership was very widespread; "Keeping the Revolution Going: Problems of Village Leadership after Land Reform," in John W. Lewis, ed., *Party Leadership and Revolutionary Power in China* (London, Cambridge University Press, 1970), pp. 239-267.
5. See, for example, Chung Wen, *China Tames Her Rivers* (Beijing, Foreign Languages Press, 1972), pp. 29-30.
6. See John P. Burns, "Chinese Peasant Interest Articulation" (PhD dissertation, Columbia University, 1979), pp. 304, 345.
7. On land disputes, see Guizhou Provincial Service, 21 July 1980, FBIS 31 July 1980, Q1; Hunan Provincial Service, 11 July 1982, P1, 2.
8. On disputes over water rights, see *People's Daily*, 4 March 1983, p. 2.
9. On woodlands disputes, see Anhui Provincial Service, 14 December 1980, FBIS 19 December 1980, O3.
10. The following case is from a report by Hainan Island Service, Haikou, 19 October 1981, FBIS, 23 October 1981, P1-2.
11. *Nanfang ribao*, Guangzhou, 5 November 1981, p. 2.
12. Hunan Provincial Service, Changsha, 5 April 1981, FBIS 8 April 1981, P14.

13. *People's Daily*, 15 March, 1979, p. 3.
14. Zhejiang Provincial Service, 27 March 1979, FBIS 28 March 1979, O6.
15. *Fujian ribao*, Fuzhou, 13 January 1982, p. 1.
16. *Ban yue tan*, Beijing, 25 December 1982, no. 24, FBIS 17 January 1983, K18.
17. *Criminal Law* in FBIS supplement, 27 July 1979, p. 49; *Guangming ribao*, 30 November 1980, pp. 1, 3.
18. *Guangming ribao*, 20 April 1981, p. 3.
19. *Fujian ribao*, Fuzhou, 13 January 1982, p. 1.
20. *Guangming ribao*, 30 April 1981, p. 3; *Ban yue tan*, 25 December 1982.
21. "Leading the deceased" was a common practice in the Yiguan Dao, as well as in more "orthodox" Buddhist religious practices. Yiguan Dao members paid a "passage price" (*dufei*) for the service of having deceased family members led to the safety of the Venerable Mother. See Kubo Noritada, "Ikkandō hokō," in *Tōyō Bunka Kenkyūjo Kiyō* 11:186 (November 1956).
22. Hunan Provincial Service, Changsha, 6 April 1980, FBIS 29 April 1980, P1.
23. Qinghai Provincial Service, Xining, 19 May 1980, FBIS 20 May 1980, T2.
24. *Nanfang ribao*, Guangzhou, 5 May 1981, p. 2, FBIS 20 May 1981, P1 (emphasis added).
25. Ibid., P2.
26. *People's Daily*, 21 April 1982, p. 8.
27. *Yangcheng wanbao*, Guangzhou, 20 May 1981, in *Inside China Mainland* (ICM), December 1981, p. 10.
28. *Nanfang ribao*, Guangzhou, 16 June 1981.
29. *People's Daily*, 29 July 1981.
30. *Ningxia ribao*, Yinchuan, 10 May 1982, p. 3, FBIS 25 May 1982, T3-4.
31. *Xinhua*, Beijing, 25 August 1981, FBIS 26 August 1981, K2. Over the next few years, further reductions in local cadres were implemented. By 1984, many teams had only 2 cadres, while brigades had 3-4 cadres.
32. Guizhou Provincial Service, Guiyang, 24 April 1981, FBIS 27 April 1981, Z1.
33. For examples of traditional Huaibei feuds, see Elizabeth J. Perry, *Rebels and Revolutionaries in North China, 1845-1945*, (Stanford, Stanford University Press, 1980), pp. 74-80.
34. Anhui Provincial Service, Hefei, 23 August 1979, FBIS 27 August 1979, O1.
35. Shandong Provincial Service, Jinan, 7 November 1979, FBIS 14 November 1979, O3.
36. For other examples of tree-cutting in orchards, see Guangdong Provincial Service, Guangzhou, 26 October 1980, FBIS 27 October

1980, P2; and Anhui Provincial Service, Heifei, 14 December 1980 FBIS 19 December 1980, O3.
37. Anhui Provincial Service, Hefei, 3 February 1980, FBIS 5 February 1980, O2-3.
38. Ibid., O3.
39. Ibid., FBIS 6 February 1980, O1-2.
40. For a discussion of traditional armed feuds, see Harry J. Lamley, "Hsieh-tou: The Pathology of Violence in Southeastern China," *Ch'ing-shih wen-t'i* 3.7:1-39 (1977).
41. The problem of local militia commanders distributing guns for private use has apparently become of some concern in recent years. In June 1980, for example, two brigade militia captains faced disciplinary action for providing a rifle and ammunition to celebrate the birthday of the grandmother of a public security agent in the commune. Shandong Provincial Service, Jinan, FBIS 13 June 1980, O1.
42. Hainan Island Service, Haikou, 21 March 1980, FBIS 25 March 1980, P7. The procurement of weapons for these inter-community feuds was apparently a lucrative business. In the spring of 1980, a police guard at the Danxian county branch of the Chinese People's Bank received 1,800 yuan for the sale of 3 guns and 40 rounds of ammunition to a production team for use in armed feuds. Two months later, the guard sold another 2 submachine guns and a rifle to a rival production team, this time netting 2,000 yuan in the transaction. Although there is no evidence that these guns were actually used in inter-community fighting, the police guard was sentenced to 15 years in prison for his entrepreneurial efforts. (*Hainan ribao*, Haikou, 18 September 1980, p. 1, FBIS 15 October 1980, P4-5.)
43. *Guangming ribao*, 20 April 1981, p. 3.
44. *People's Daily*, 15 July 1981, pp. 1, 4.
45. *Nanfang ribao*, Guangzhou, 5 May 1981, p. 2.
46. *Shaanxi ribao*, Xi'an, 9 November 1980, FBIS 13 November 1980, T1.
47. *Guangming ribao*, 20 April 1981, 3; *Nanfang ribao*, Guangzhou, 5 May 1981, p. 2.
48. *People's Daily*, 6 January 1982, p. 4.
49. *Ban yue tan*, 25 December 1981.
50. *People's Daily*, 12 July 1982; also cited in *Wen hui bao*, Shanghai, 11 October 1982, p. 3.
51. *Xinhua*, Beijing, 2 September 1983, FBIS 6 September 1983, K12.
52. Certainly rural mobility has been on the increase in recent years. See, for example, *Xinhua*, Hangzhou, 29 February 1984, FBIS 29 February 1984, O5-6; *People's Daily*, 16 February 1984, p. 3.
53. For Europe, see especially Charles Tilly, Louise Tilly, and Richard Tilly, *The Rebellious Century, 1830-1930* (Cambridge, Harvard Univeristy Press, 1975). For the Third World, see in particular James C. Scott, *The Moral Economy of the Peasant* (New Haven, Yale Uni-

versity Press, 1976). The threat of class-based peasant protest was seen in China in the winter of 1979 when thousands of peasants demonstrated in Beijing, carrying banners and shouting slogans calling for an end to hunger and injustice. On 14 January, a group of 100 or so angry peasants yelling "We're tired of being hungry" and "Down with oppression" tried to storm the residence of then Party Chairman Hua Guofeng in an unsuccessful attempt to present their grievances directly to him. (See the reports by AFP correspondents Georges Biannic and Francis Deron, FBIS 9 January 1979, E1, 15 January 1979, E1, 22 January 1979, E1, 23 January 1979, E2, 25 January 1979, E2).

Chapter 8. The Politics of Industrial Reform, by Susan L. Shirk

1. A. Doak Barnett, *Cadres, Bureaucracy, and Political Power in China* (New York, Columbia University Press, 1967), p. 6.
2. Jerry F. Hough, *The Soviet Union and Social Science Theory* (Cambridge, Harvard University Press, 1977), p. 47.
3. Ibid., p. 83.
4. For discussions of the "cycle of reform" in the Soviet Union and Eastern Europe, see Gertrude E. Schroeder, "The Soviet Economy on a Treadmill of 'Reforms'," U.S. Congress Joint Economic Committee, *Soviet Economy in a Time of Change* (Washington, U.S. Government Printing Office, 1979), I, 312-340; and Ronald Amann, "Industrial Innovation in the Soviet Union: Methodological Perspectives and Conclusions," in Ronald Amman and Julian Cooper, eds., *Industrial Innovation in the Soviet Union* (New Haven, Yale University Press, 1982), pp. 30-37.
5. For a good description of the basic features of the Chinese economic system, see Benjamin Ward, "The Chinese Approach to Economic Development," in Robert F. Dernberger, ed, *China's Development Experience in Comparative Perspective* (Cambridge, Harvard University Press, 1980), pp. 91-119.
6. During the 1949-1978 period, the gross value product of heavy industry multiplied 90.6 times, while agriculture and light industry rose only 2.4-fold and 19.8-fold respectively; Dong Furen, "The Chinese Economy in the Process of Great Transformation," in George C. Wang, ed, *Economic Reform in the P.R.C.* (Boulder, Westview Press, 1982), p. 136.
7. The argument that the command economy inevitably generates an expansion drive which produces shortages, and that these shortages motivate local self-sufficiency, was made by the Hungarian economist Janos Kornoi and has been elegantly applied to China by Christine Wong, "The Economics of Shortage and the Problem of Post-Mao Reforms in China" (unpublished paper, 1982).

8. Nicholas R. Lardy, *Economic Growth and Distribution in China* (New York, Cambridge University Press, 1979).
9. "Yao Yilin's report on the 1983 economic plan at the First Session of the Sixth National People's Congress," *Xinhua*, 24 June 1983, FBIS 24 June 1983, K6.
10. *People's Daily*, 9 May 1983, FBIS 26 May 1983, K15.
11. For an excellent discussion of the effect of the informal relations of the factory on the implementation of wage and bonus reforms, see Andrew G. Walder, "Wage Reform and the Web of Factory Interests," in David M. Lampton, ed., *Policy Implementation in Post-Mao China* (forthcoming).
12. This is the argument Charles F. Sabel and David Stark make for Soviet and East European societies in their stimulating article, "Planning, Politics, and Shop-Floor Power: Hidden Forms of Bargaining in Soviet-Imposed State–Socialist Societies," *Politics and Society* 11.4:439–475 (1982).
13. For an excellent discussion of the weakening central control over resource flows, see Christine Wong's chapter in this volume.
14. Jack Craig, Jim Lewek, and Gordon Cole, "A Survey of China's Machine-Building Industry," U.S. Congress, Joint Economic Committee, *China Economy Post-Mao* (Washington, U.S. Government Printing Office, 1978), p. 291.
15. Wong's chapter in this volume; also see *Beijing Review*, 18 April 1983, pp. 24–25.
16. *China Business Review*, September–October 1982, p. 26.
17. Dorothy J. Solinger, "The Fifth National People's Congress and the Process of Policymaking: Reform, Readjustment, and the Opposition," *Issues and Studies* 18.8:63–106 (August 1982).
18. Alec Nove, *The Economics of Feasible Socialism* (London, Allen and Unwin, 1983), p. 79.
19. On the relative political influence of various central ministries in the economic policy process, see Michel Oksenberg, "Economic Policy-Making in China," *The China Quarterly*, no. 90 (June 1982).
20. Nicholas R. Lardy, "Economic Planning in the People's Republic of China: Central-Provincial Relations," in U.S. Congress, Joint Economic Committee, *China: A Reassessment of the Economy* (Washington, U.S. Government Printing Office, 1975), pp. 94–115. For a reassessment of the impact of Chinese policies on regional inequalities, see Suzanne Paine, "Spacial Aspects of Chinese Development: Issues, Outcomes and Policies, 1949–79," *The Journal of Development Studies* 17.2:135–195 (1981).
21. Charles Robert Roll, Jr., and Kung-chia Yeh, "Balance in Coastal and Inland Industrial Development," U.S. Congress Joint Economic Committee *China: A Reassessment of the Economy* (Washington, U.S. Government Printing Office, 1975), pp. 81–93.

22. Audrey Donnithorne, *The Budget and the Plan in China: Central-Local Economic Relations* (Canberra, Australian National University, 1972); and Donnithorne, *Centre-Provincial Eonomic Relations in China* (Canberra, Australian National University, 1981).
23. *People's Daily*, 31 December 1982, FBIS 4 January 1983, P22.
24. *China Business Review*, September–October 1982, P21.
25. *Xinhua*, 29 November 1982, FBIS 3 December 1982, I3.
26. *Xinhua*, 16 May 1983, FBIS 17 May 1983, P3.
27. *Ming pao* (Hong Kong), 6 July 1982, FBIS 6 July 1982, W3.
28. During 1980–1982, over one-third of the cost of capital construction in the Shenzhen SEZ came from foreign investors (*Nanfang ribao*, 14 June 1982, FBIS 24 June 1982, P2).
29. Zhang Peiji, "Stick to Open Policy and Expand Foreign Trade," *Economic Reporter* (Hong Kong), May 1982.
30. *CZ*, no. 7 (5 July 1982), pp. 8-9. JPRS, 82018, 19 October 1982, p. 19.
31. In 1983, Shanghai's autonomy was expanded and the terms she could offer foreign companies were improved to attract more foreign capital to China. *Xinhua*, 16 May 1983, FBIS 17 May 1983, O3.
32. Xue Muqiao, "Economic Management in a Socialist Country," in George C. Wang, ed., *Economic Reform in the P.R.C.* (Boulder, Westview Press, 1982), p. 33. A similar pattern has been found in Japan by T. J. Pempel and Keiichi Tsunekawa, "Corporatism Without Labor? The Japanese Anomaly," in Philippe C. Schmitter and Gerhard Lehmbruch, eds., *Trends Toward Corporatist Intermediation* (Beverly Hills, Sage Publications, 1979), p. 269. They point out that "economic sectors that see themselves as feeble in the face of extensive international competition" tend to seek protection from the state in the form of corporatism, while, for sectors that can compete successfully, "the protections corporatism promises are not worth the sacrifices it demands of the strong."
33. *Shanxi ribao*, 2 August 1982, FBIS 13 August 1982, R3-5; Shanxi Provincial Service, 11 December 1982, FBIS 15 December 1982, R2-3.
34. For a fuller discussion of this controversy, see Susan L. Shirk, "The Domestic Political Dimensions of China's Foreign Economic Relations," in Samuel S. Kim, ed., *Chinese Foreign Policy in the 1980s* (Boulder, Westview Press, forthcoming).
35. *Shijie jingji daobao*, 1 November 1982, p. 4; *People's Daily*, 27 April 1980; *China Daily*, 22 July 1982; *Shijie jingji daobao*, 28 June 1982, FBIS 23 July 1982, O3; *Liaowang*, 20 June 1982, FBIS 3 August 1982, K17; *People's Daily*, 2 August 1982, FBIS 10 August 1982, K10; *People's Daily*, 16 August 1982, FBIS 26 August 1982, K6; *China Daily*, 22 January 1983.
36. *South China Morning Post* (Hong Kong), 14 September 1982, FBIS 14 September 1982, W6-9; *China Business Review*, September–October 1982, p. 4.

37. *Jingji ribao*, 3 January 1983, FBIS 18 January 1982, K15-16, K23-24; *Xinhua*, 12 January 1983, FBIS 18 January 1982, K8.
38. *People's Daily*, 20 September 1982, p. 5.
39. MOFERT has been more successful at getting control over exports by imposing export licenses and taxes. *Zhongguo xinwen she*, 23 July 1982, FBIS 27 July 1982, P1; Japan External Trade Organization (JETRO), *China Newsletter*, no. 41 (November-December 1982), p. 18; *Caimao jingji*, no. 6 (15 June 1982), JPRS, 81938, 8 November 1982.
40. Franz Schurmann, *Ideology and Organization in Communist China* (Berkeley, University of California Press, 1968), p. 197.
41. The center may prefer to build railroads, while the localities favor roads, but this is a political choice having to do with the locus of control over transportation and not a purely economic one. Samuel Popkin, in personal communication, made this point to me.

Chapter 9. False Starts and Second Wind: Financial Reforms in China's Industrial System, by Barry Naughton

Grateful acknowledgment is made to William Byrd, Nicholas Lardy, Christine Wong, and Wu Jinglian.

1. The most comprehensive statement of this point of view is in Liu Guoguang and Wang Ruisun, *Zhongguo de jingji tizhi gaige* (Reform of China's economic system; Beijing, 1982), especially p. 9.
2. Cf. Jiang Yiwei, "Qiye Benwei Lun," *Zhongguo shehui kexue* (Chinese social science) no. 1 (1980). This was reprinted in English under the title, "Theory of an Enterprise Based Economy."
3. "Every country's situation is different, and economic reforms take innumerable forms. But it's worth paying attention to the fact that every (socialist) country makes the adjustment of the financial authority of the enterprise the central link in the reform process." Yang Jianbai, "Guanyu guomin jingji de tongyi lingdao he qiye de zizhuquan wenti" (On the problem of unified leadership of the national economy and enterprise autonomy), in Liu Guoguang, ed., *Guomin jingji guanli tizhi gaige de ruogan lilun wenti* (Several theoretical issues on the reform of the economic management system of the national economy; Beijing, 1980), p. 63.
4. This recognition has even attained "official" status, since it was included in a document of the central government; *State Council Bulletin* (1981) pp. 439-440.
5. This account generalizes a somewhat more complex reality. In particular, there were a variety of slightly different methods for calculating profit retention. See Bruce Reynolds, "Reform in Chinese Industrial Management: An Empirical Report," in U.S. Congress Joint Economic Committee, ed., *China Under the Four Modernizations*, (Washington, U.S. Government Printing Office, 1982) I, 119-137. The initial reform

documents of 13 July 1979 were phrased in very general terms, with substantial scope for local variation.
6. Liu Guoguang, "Jingji guanli tizhi gaige de tuogan zhongyao wenti" (Several important questions relating to the reform of the economic management system), in Liu Guoguang, ed., *Guomin jingji. . . .*, p. 4.
7. Tian Chunsheng, *JJGL* 2 (1979), translated (poorly) in JPRS E.A. 3 August 1979; Zuo Chuntai and Xiao Jie, "An Exploration of the Reform of the Drawing, Use, and Management of Basic Depreciation Funds," *Caimao jingji* 5:4-8 (1983).
8. See *Guojia yusuan* (The state budget; Beijing, 1980), pp. 48-53.
9. See sources cited in note 7. Also Liang Wensen and Tian Jianghai, "Wo guo jiben zhejiu jijin guanli zhidu de yanbian" (The evolution of the system of managing basic depreciation funds in China), *Caiwu yu kuaiji* 9:25 (1979). These funds were not necessarily under the actual control of the enterprises; more typically they were effectively controlled by local governments, which readily used the majority of them for new construction, rather than replacement of aging fixed assets. The articles cited in this note make this point quite clear.
10. Zhu Fulin, "Some Opinions on the Reform of Our Nation's Budgetary System in the Past Several Years," *CZ* 2:17 (1983).
11. "Commentator," *Caiwu yu kuaiji* 11:3 (1979); Ling Chen, "State-run Enterprises Ought to Actively and Systematically Try out the Profit Retention System," *JJGL* 12:15-17 (1979); Ren Tao et al., "Investigation Report: Enterprises in Sichuan Province Acquire Greater Independence," *Social Sciences in China* 1:209-210 (1980).
12. See, e.g., Ma Hong, "Transform the Economic Management System," *Hongqi* 10:50-59 (1979), translated in JPRS #74680.
13. *Jingjixue dongtai* 11:1-5 (1979).
14. *Caiwu yu kuaiji* 11:1 (1979).
15. Ling Chen, "State-Run Enterprises," p. 17. Ren Tao et al., pp. 214-215.
16. *Caiwu yu kuaiji* 3:16, 18 (1980).
17. *Beijing Review*, 18 August 1980, p. 3; *People's Daily*, 2 January 1981, p. 1.
18. *Gongren ribao*, 11 April 1980, p. 1, tr. in JPRS Ec. #64.
19. Ling Chen, p. 17; *Caiwu yu kuaiji* 5:29-30 (1981). The five provinces were Shanghai, Hebei, Henan, Shandong, and Jiangsu.
20. Although the provision that called for raising depreciation rates was scaled back after 1979, and further raises were quite limited. See Zuo Chuntai and Xiao Jie, p. 4.
21. *People's Daily*, 20 August 1980, p. 1.
22. *Gongye jingji guanli congkan* 2:6 (1981).
23. Currency in circulation increased 30% during 1980, but cash income of the population expanded by an equal amount, due to increased earnings and increased monetization of the rural economy. On this

score, at least, the alarm felt by central planners would appear to be misplaced.
24. Bank loans are repaid out of profits *before* the enterprise's retention share is computed. Moreover, tax remittances are frequently permitted to be diverted to loan repayments. See Dai Yuanchen, "An Exploration of the Problem of the Source of Loan Repayment Funds," *CZ* 2:22-23 (1982).
25. The document on equipment loans in *State Council Bulletin* (1981), pp. 380-384, makes reference to the process of "unfreezing" loans. These events and chronology were described at enterprise interviews conducted in Wuhan, April 1982.
26. Reynolds, p. 136; T. Pairrault, *World Development* 8:642 (1983).
27. Document of 26 January 1981 in *State Council Bulletin* (1981), p. 37. The situation in which higher authorities are unable to secure enterprise acceptance of profit targets (*jihua buneng luoshi*) until a successful process of profit contract bargaining is described in a number of sources. See Tian Xinyi, "Actively Assist the Enterprises to Uncover Latent Potential," *CZ* 11:1-4 (1981), in Fushun; Yu Youhai, "Don't Throw Out the Baby with the Bathwater," *Gongye jingji guanli congkan* 7:12-16 (1982), in Shandong and Shanghai; Wang Changxiao, "Underwriting Profits, Sharing Additional Income," *Hubei ribao* 1 August 1981, p. 1, tr. in JPRS Ec. #175, for Wuhan.
28. Yu Youhai, p. 12.
29. Wang Changxiao, p. 1; Sun Xuewen, "We Must Correctly Carry Out Enterprise Responsibility Systems," *Caimao jingji* 2:30-34 (1982).
30. Sun Xuewen, p. 31.
31. Yu Youhai, pp. 12-13; Ren Tao, "In the Final Analysis, What Influence Does Industrial Management Reform have on Budgetary Revenues?" *JJGL* 11:31-34 (1982).
32. People's Bank of China, Nanjing Branch, "Explore a New Path for Transformation Loans in Old Enterprises," *Zhongguo jinrong* 18:12, 21 (1982).
33. Commentator, "Actively Explore and Experiment With Industrial Production Responsibility Systems," *People's Daily*, 5 August 1981, p. 2.
34. Gu Fuwen, *CZ* 8:26-27 (1982).
35. Sun Xuewen, p. 33. This may not be entirely fair to the profit-contract system. We do not know the extent to which areas that implemented the profit-contract system less vigorously failed to fulfill *their* plans. Also, there is some evidence that, in the case of Shandong, the cause of underfulfillment was primarily the poor performance in the first quarter before the implementation of the profit-contract system. See Yu Youhai, p. 13.
36. And these arguments were accepted by their superiors. See "Profit Contracting: Immediate Efficacy," *People's Daily*, 5 August 1981, p. 2.

37. This chain is described with particular clarity for Jinan in Wei Jianyi, "The Economic Responsibility System is Effective Wherever it's Used," *Qiye guanli* 5:2 (1981). Moreover, in this case the chain extends into the enterprise to the section, workshop, and even individual worker.
38. See, for instance, Wang Renzhi, Gui Shiyong, and Xu Jingan, "Restructuring the Economic Management System," *Hongqi* no. 5 (1980).
39. *State Council Bulletin* (1981), pp. 439-452.
40. There have been numerous descriptions of the Capital Steel model. See, for instance, *People's Daily*, 30 November 1981, and *JJGL* no. 3 (1982) and no. 4 (1982).
41. Cf. Gui Shiyong Shehui, *Zhehui zhuyi guoying gongye qiye zerenzhi* (The responsibility system in socialist state-run industrial enterprises; Beijing, 1964).
42. The former criticism was more often voiced privately, since it is considered bad form to criticize a national model publicly. As for the second view, after the resurgence of reform in early 1983, it became the official view; as a representative from the State Economic Commission declared, "The crux of the experience of Capital Steel is reform." Beijing Domestic Service of 30 January 1983, translated in JPRS Ec. #315.
43. See Sun Xuewen; Yu Youhai.
44. Compare the progressive scaling down of goals for enterprise rectification reflected in Yuan Baohua, *JJGL* 9:16-18 (1982); Ceng Zhi, *Qiye guanli* 1:7 (1983); and *People's Daily*, 21 September 1983, p. 1.
45. The Wuhan Iron and Steel Company was directed to shift to the profit-contract system in September 1982 as part of the enterprise rectification program (Wuhan interview, 16 September 1982). This was almost certainly canceled when the focus shifted to tax-for-profit, which Wuhan Steel had been using since early 1981, for its experience was discussed (as a continuing experiment) in Zheng Lilin, "An Analysis of the Situation in a Portion of the Provinces Implementing Tax-for-Profits Experiments," *Jingji lilun yu jingji guanli* 2:1-5 (1983).
46. *Jingji daobao* (H.K.), 28 January 1981, tr. in JPRS Ec. #123; Niu Licheng and Yu Zhizhong, "An Investigation of Small-scale State Factories in Guanghua County Shifting to Income Taxes and Responsibility for Profits and Losses," *CZ* 11:16-18 (1891); and Zheng Lilin.
47. *State Council Bulletin* (1981), pp. 439-452, and (1982), pp. 28-32.
48. Most of the enterprises I visited in Wuhan in spring 1982 had not implemented fixed-capital charges (Wuhan Iron and Steel being the main exception).
49. Zheng Lilin; Gu Zongcheng, "Some Problems Related to Profit-Loss Responsible Enterprises," *JJGL* 12:27-28 (1980).
50. Gong Zhi, "Implement Compensated Use, Raise the Efficiency of Capital," *CZ* 2:12 (1982).
51. Shanghai Municipal Finance Bureau, "Respect and Bring Into Play Accounting Work in order to Raise Economic Efficiency," *Caiwu yu*

kuaiji 2:1-4 (1982); Wu Cuilan, "Several Problems Regarding the Compensated Use of Capital," *Caiwu yu kuaiji* 4:7-9 (1981).
52. "Tentative Methods for Changing Profit to Tax in State-Run Enterprises," *Xinhua yuebao* 4:116-118 (1982).
53. Xinhua Domestic Service of 8 February 1983, tr. in JPRS Ec. #319.
54. *State Council Bulletin* (1981), p. 271. This may understate the total somewhat, since the results are reported only for the 5,777 enterprises subordinate to local governments. Thus, several hundred enterprises subordinate to central government ministries are omitted. But this could not change the basic picture sketched in the text.
55. Lin Senmu and Zhou Shulian, *Hongqi* 3:9-13 (1981).
56. *Gongye jingji guanli congkan* 2:8 (1982).
57. Kang Zhixin, "Capital Construction," *1982 JJNJ*, p. V-298.
58. *1983 Tongji zhaiyao*, pp. 60-62; "Report on Fulfillment of the 1982 Plan," *Xinhua yuebao* no. 4 (1983). I have discussed this problem at some length in "The Decline of Central Control Over Investment in Post-Mao China," forthcoming in a volume edited by M. D. Lampton.
59. This viewpoint is expressed most consistently in the various essays in Jingji Yanjiu Editorial Board, eds., *Guanyu woguo jingji guanli tizhi gaige de tantao* (An exploration of reform of the economic management systems in China; Jinan, 1980), most forcefully in the essay by Sun Yefang.
60. Janos Kornai, *Economics of Shortage* (Amsterdam, North Holland, 1980).
61. *People's Daily*, 18 September 1983, p. 2.
62. Naughton, "Decline of Central Control."
63. *1983 Tongji zhaiyao*, p. 64.
64. Zhao Xiangyang, "We Must Pay Attention to the Guidance of Planning Even More, Now that Enterprises have Increased Autonomy," *Gongye jingji guanli congkan* 3:31 (1980).
65. Zhang Pinqian and Xiao Liang, "Since the Means of Production Have Entered the Marketplace," *Zhongguo shehui kexue* no. 5 (1980).
66. Yu Youhai, p. 14.
67. *State Council Bulletin* (1981), pp. 439-452.
68. Beijing Radio, 7 September 1983, tr. in FBIS 13 September 1983, K-4.
69. Song Ping, cited in FBIS 20 July 1983, K-15; *People's Daily*, 16 July 1983, p. 1.

Chapter 10. Material Allocation and Decentralization: Impact of the Local Sector on Industrial Reform, by Christine Wong

Research for this chapter was carried out under grants from the Committee for Scholarly Communications with the People's Republic of China, Social Science Research Council, and a postdoctoral fellowship at the

Fairbank Center for East Asian Research, Harvard University. Grateful acknowledgment is due to William Byrd, Dwight Perkins, Elizabeth Perry, and Lee Travers for helpful comments on an earlier draft.

1. One report estimates that bottlenecks in fuel, electricity, and transport supplies caused 20-30% of industrial capacity to be idled during the mid-late 1970s; Ma Hong and Sun Shangqing, eds., *Zhongguo jingji jiegou wenti yanjiu* (Research on the problems of China's economic structure; Beijing, 1981).
2. Li Kaixin, "Concentrate Materials to Ensure Keypoint Construction," *Hongqi* 17:17 (1983). Also, Andrew Walder cites an informant on factory management: "What good is money? Everybody has money. It is the goods which are valuable"; "Industrial Reform in China: The Human Dimension," in Ronald A. Morse, ed., *The Limits of Reform in China* (Boulder, Westview Press, 1983), p. 47.
3. Li Kaixin, p. 16.
4. Yao Yilin, "Report on the 1983 Plan for National Economic and Social Development (Excerpts)," *Beijing Review,* 11 July 1983, p. II.
5. Interview in Beijing, July 1982; and Li Jingwen, "Problems Concerning the Nature and Management of the Circulation of Producers' Goods," in Liu Guoguang, ed., *Guomin jingji guanli tizhi gaige de ruogan lilun wenti* (Some theoretical problems of the reform of the national economic system; Beijing, 1981).
6. Zhang Shushan, "Some Views about Reforms in Commodity Controls," *JJGL* 8:15-18 (1979), translated in JPRS *Economic Affairs* (E.A.) 34:16.
7. Li Kaixin, "Management of China's Materials and Resources," *1981 JJNJ* (Beijing, 1982) p. IV-124.
8. Li Jingwen, p. 203.
9. Ibid.
10. Unless otherwise noted, information in this paragraph comes from Liu Guoguang and Wang Ruisun, *Zhongguo de jingji tizhi gaige* (Reform of China's economic system; Beijing, February, 1982), pp. 3-10.
11. See, for example, Bruce Reynolds, "Reform in China's Industrial Management: an Empirical Report," in Joint Economic Committee, U.S. Congress, *China Under the Four Modernizations* (Washington, U.S. Government Printing Office, 1982), p. 130.
12. With the introduction of joint ownership and multiple sources of investment finance, this became very complicated. In Kaiping county, Guangdong, I was told that county fertilizer plants built entirely with county monies kept all their output for the counties' use, while those built partially with provincial funding had to "remit" part of their output for provincial allocation (fieldwork in Guangdong province, 1978).
13. Liu Guoguang and Wang Ruisun, pp. 8-9.
14. Audrey Donnithorne, *China's Economic System* (London, Allen and Unwin, 1967), p. 461; Dwight Perkins, *Market Control and Planning in China* (Cambridge, Harvard University Press, 1966), pp. 107-116.

15. Herbert S. Levine, "The Centralized Planning of Supply in Soviet Industry," JEC, *Comparisons of the U.S. and Soviet Economies* (Washington, U.S. Government Printing Office, 1959) p. 155. However, it should be noted that Chinese classifications of materials involve broader categories than the Soviet system, so that the numbers are not strictly comparable.
16. Sun Shangqing and Chen Shengchang, *Zhongguo de chanye jiegou* (The structure of Chinese industry; Beijing, May 1982), p. 56.
17. Zhang Shushan, p. 16.
18. Li Jingwen, p. 204.
19. Zhang Shushan, p. 16.
20. During the CR period, restrictions included a prohibition on barter involving consumer goods. Strictly speaking, only extra-plan and locally allocated goods were allowed to enter this trade, and there were limits on distances shipped, etc.
21. Wang Gengjin and Zhu Rongji, "Whither Commune and Brigade Industry?" *JJGL* 3:21 (1979).
22. This is especially true at the county level, where county economic bureaux assumed control over the supply and marketing, not only of county enterprises, but also for commune and brigade enterprises.
23. Yu Xiaogu, "Implement the Merger of Planned and Market Adjustment to Liven up Material Circulation," *JJGL* 2:4 (1980).
24. Xing Hua, "The Frequent Changes in the Fiscal System During the Ten Years of Chaos, Part Two," *CZ* 9:8 (1983); Liu Guogang and Wang Ruisun, p. 7.
25. Sun Shangqing and Chen Shengchang, p. 9.
26. Du Zhenbiao, "The Correct Handling of the Problem of Losses in Local Coal Mines in Jiangnan," *JJGL* 8:11 (1980).
27. For example, Liaoning's expenditures on nonproductive investment fell from 19% of total during the FFYP to 9.9% in the 3FYP and 12.6% in the 4FYP. Shanxi's expenditures on nonproductive investment averaged 14.5% of total during 1949-1978, but fell to a low of 8.6% during the 3FYP. Huang Rongsheng, "On the Comprehensive Balancing of Local Economies," in Liu Guoguang.
28. Severe shortages during the 1970s reportedly caused a "three eights" system to be implemented, whereby enterprises were allocated only 0.8 of amounts of materials requested, orders were accepted for only 0.8 of the allocated amounts, and deliveries covered only 0.8 of the orders. See Li Jingwen, p. 253.
29. Ding Hua, "A Basic Cause of Poor Results of Investment Lies in the Economic Management System—an Investigation of the Iron and Steel Works of the Municipality of Huangshi," *JJYJ* no. 3, 1981, JPRS E.A. 139:40-52.
30. See Table 26. A separate account reported that materials decontrolled included 78 metallurgical products, 144 machinery and electrical products, etc.; 1981 *JJNJ* p. IV-125. The discrepancies in numbers

come from slight differences in categorizations that are as yet unclear to me. There was talk of reasserting ministerial control over some 600 materials in 1978-1979. It is believed that this was not implemented, although some reports referred to the higher number of materials under central control. For example, see Li Jingwen, p. 204.
31. 1980 inventory sales included 1.28 million tons of steel, equal to 40.1% of excess stocks, and 3.35 billion yuan of machinery and electrical products, 25% of excess stocks; 1981 *JJNJ* p. IV-126.
32. *Wuzi guanli* (Materials management) 4:1 (1983).
33. Ibid., pp. 13-14.
34. Ibid., p. 8.
35. For studies of the development of local industries during the CR, see Carl Riskin, "China's Rural Industry: Self-Reliant Systems or Independent Kingdoms?" *China Quarterly* 73:77-98 (March 1978); and Christine Wong, "Rural Industrialization in the People's Republic of China: Lessons from the Cultural Revolution Decade," in *China Under the Four Modernizations*.
36. The impression I obtained from field work in 1981 and 1982 was that county economic committees and finance departments were playing the primary role in investment decisions. They were also very much involved in the day-to-day operations of enterprises. Numerous articles in *CZ* and other journals glorifying the work of finance-department cadres give details of the active role these cadres play in obtaining supplies and markets for local enterprises, along with helping them to straighten out their management and accounting procedures.
37. Interview, Beijing, August 1982. See also Audrey Donnithorne, "New Light on Central-Provincial Relations," *The Australian Journal of Chinese Affairs* no. 10, 1983.
38. Yang Xuguang, Xu Changzhong, and Zhang Guangyi, "Adopt Economic Measures, Reform Coal Supplies," *Zhongguo caimao bao* (Bulletin of Chinese finance and trade), 16 October 1982, JPRS E.A. 295:28.
39. *Shanxi ribao*, 19 October 1982, p. 2; JPRS E.A. 323:80.
40. Xue Muqiao, "Some Theoretical Problems Concerning the Reform of the Economic Structure"—Speech Delivered at the Symposium on Economic Reform, 25 September 1982; *JJYJ* 1:5 (1983).
41. Tao Fei et al., "An Investigation of the Consumer Goods Market in Tianjin," *JJYJ* 2:68 (1983).
42. Xue Muqiao, p. 5.
43. Field work in Jiangsu province, June 1982.
44. Yang Xuguang et al., p. 28.
45. Li Jingwen and Liu Binzhe, "Strive to Accelerate Development of Our Plate Glass Industry," *JJGL* 1:28 (1982).
46. Field work in Guangdong province, July 1982.
47. Li Jingwen, "Attention Should Be Paid to Circulation," *Shijie jingji dabao* (Shanghai, Bulletin of world economy), 27 September 1982, p. 4; in JPRS E.A. 300:23.
48. Xue Muqiao, p. 5.

49. Liu Zhiyuan, Chao Gangling, Cao Yang, Ma Hui, and Chen Qijie, "The Causes of and Solution for the Present Overstock of Chemical Fabrics," *Caijing yanjiu* (Study of finance and economics) 5:43-48 (1982), in JPRS E.A. 305:2.
50. Xue Muqiao, p. 5.
51. In 1983, tobacco and liquor taxes were reassigned to central revenues. See *CZ* 7:37-38 (1983).
52. Xue Muqiao, p. 5.
53. Many reports of this can be found in the press. For example, an NCNA report said that some localities "are unwilling to ship out their raw materials to support technically advanced industrial cities according to state plan. . ." 12 November 1980, in BBC, *Summary of World Broadcasts*, 6581/c/2.
54. Xue Muqiao, p. 5.
55. NCNA (Chinese), 28 March 1983, in BBC W1233/A/3.
56. Hao Shengde, "From the Changing Costs of Production in Commune and Brigade Coal Mines to the Necessity of Adjusting Prices and Taxes," *Jingji wenti* (Economic problems) 5:41-42 (1982).
57. For examples of local protectionism, see Liu Guoguang and Wang Ruisun, pp. 16-17, 57; and *JJGL* 2:57, 1979.
58. Beijing Home Service, 6 October 1979, in BBC, W1055/A/14.
59. *Tongji nianjian* (Statistical yearbook), 1983, p. 245; NCNA 7 May 1983, JPRS, E.A. 345:69; and NCNA 5 December 1978, in BBC, W1013/A/12.
60. *Wuzi guanli* 1:12 (1983).
61. Li Kaixin, p. 16.
62. For an argument of the importance of external economies and linkages in developing economies, see Charles K. Wilber, "Economic Development, Central Planning and Allocative Efficiency," in *Yearbook of East European Economies, 1971*, pp. 221-243.

INDEX

Above-quota prices. *See* Prices, above-quota
Above-quota quotas. *See* Quotas, above-quota
Absenteeism, worker, 4
Agriculture, 31–192; and reform, 1–11, 14–15, 22–24, 53–61, 176; and Hua's ten-year plan, 9; and collectives, 11–12, 72; and production responsibility systems, 34–53; products, and rural market trade, 45; and size of farm unit, 86; and government investment, 129; and Single-Child Family policy, 131–156;
Agro-industrial-commercial enterprises, 53
Agro-technician (non-leading cadre), 163, 171
Anhui province: and agricultural production responsibility systems, 14, 40, 47–48, 120, 158, 213; and private trade, 44; and local leadership, 162–165; and rural violence, 186–192
Animals, draught, and household contracting, 47–48
Anshan Iron and Steel Company: and local leadership, 208; and local industry, 256
Assembly industries, and local investment, 14
Atheism, and Communist Party, 183

Bai Doncai, Governor of Jiangxi, and production responsibility systems, 158–159
Bank loans, repayable: and industry, 12, 200, 232; and specialized households, agricultural, 46; and rural income, 114, 122
Banks: and local leadership, 24, 170; and production responsibility systems, 122; and industry, 200, 226, 231–232
baochan daohu, 35–36, 168
baochan daozu, lianchan jichou, 35–36
baogan daohu, da baogan, 35–36, 49, 64–82, 166, 168
Baoshan steel complex, and Hua's ten-year plan, 9
Bargaining, 238, 243, 245

Beijing, and economic reforms, 210, 213, 230, 236, 242. *See also* Centralization; Government
Birth control. *See* One-Child Family policy; Planning, family
Black market, and foreign exchange, 212
Bonuses, workers': and reform, 9, 10, 201-202; and Communist Party, 17; and household contracting systems, 36-38; and rural income, 114; and local leaders, 170-171; and industrial reform, 227, 235. *See also* Incentives
Brigade, production: and education, 20; and collective ownership, 33, 51; and technical responsibility systems, 48-49; and new economic combinations, 50; and factories, 72; and rural income, 112-113, 116-117, 125-126; and production responsibility systems, 160-161; cadres, 165-173; and rural violence, 177-192
Bureaux, industrial. *See* Ministries

Cadres. *See* Leadership, local; Leadership, state
Capital, fixed, working: and interest charges, 12, 226, 231, 241-244, 252; and investment, 197, 200; and interprovincial cooperation, 215
Capital-construction projects: and commune labor, 125; and local fixed investments, 229, 255
Capitalism: and reform, 1; and Deng Xiaoping, 7; and households, prereform, 64; and incentives, 64-65; and agricultural property rights, 81; Overseas Chinese, 211
Capital Steel Company, 239-241
Careers, and local leadership, 161-173
Category I, II, III: and agricultural products, 85-109; and industrial products, 253-278
Cement, 242, 250, 255, 260-278; and supply shortages, 201
Centralization, 197-221; and Hua's ten-year plan, 9; and Cultural Revolution, 13-14; and agricultural production responsibility systems, 15; and "Communist coalition," 16; and Dazhai collectivism, 32; and industrial reform, 227-228. *See also entries beginning* Decentralization; Planning
Chen Muhua, MOFERT chairman, and foreign trade, 217
Chen Yun, economist: and Single-Child Family policy, 142; and economic reform, 254
Children: and key-point household labor, 46; and collectivization, 80; and Single-Child Family policy, 131-156
China International Trust and Investment Corporation, and joint ventures, 218
Chinese Academy of Social Sciences, 4
Cigarette industry: and rural income, 116; and prices, 199, 249, 272; and taxes, 272-273
Cities, coastal, and industrial reforms, 210-211
Coal industry: and reforms, 201-210, 249, 260-278; and foreign trade, 206
Collective ownership: of land, 23-24; Maoist model (Dazhai system), 32-34; and agricultural reform, 37, 48-61, 63-82, 111-130; and private exchange, 104; and local leadership, 167-173; and rural violence, 177-192; and economy, 197; factories, and industrial reforms, 202
Command economy, Soviet-style, 197-204. *See also* Centralization; Government
Commerce. *See* Trade
Commune: and education, 20; and collective ownership, 33; and agriculture, 48-50; and government, 50-53; and reform, 51-53, 64, 68-69; in Jiangsu, 77; and rural income, 112-113, 116-117; and marketing area, 178; and rural violence, 178-192
Communist Party, Chinese: Third Plenum, Eleventh Congress, and reform, 1-2, 10-27, 63, 111, 115, 132, 161, 178-179; reform of ranks, 6, 16-19, 252; Central Committee, 8, 38-43, 250-251; Central Discipline Inspection Commission, 17; Fifth Plenum, 17; Party Secretariat, 17; Twelfth Party Congress, 17, 157-158; National People's Congress, debate at, 19; and commune reorganization, 51; and rural violence, 175-192; and economic reform, 195-198; and coastal provinces, 210. *See also* Government
"Communist coalition" and industrial reforms, 16, 199, 204
Comparative advantage. *See* Regions, and trade
Compensation. *See* Remuneration, Wages
Competition: and industrial reform, 203-221, 275; international, and heavy industry, 206

Concrete. *See* Cement
Construction, materials, 201, 203, 219–220, 249, 269–278
Consumption, agricultural, rural: and grain quotas, 88; and state procurement prices, 96–103; and income, 111–112, 125–127, 165
Consumption, industrial, urban: and Cultural Revolution, 4; and rural income, 126–127; and industrial reform, 247
Consumer goods. *See* Goods, consumer
Contracts, foreign: and Hua's ten-year plan, 9; and investment, 246
Contracts, production: and legal reform, 19; and household contracting systems, 34–61, 63–82; and collectivization, 72, 76–82, 128; and family planning, 151; and foreign countries, 246
Contracts, and profit, 11–12, 24, 233–240
Cooperatives, agricultural, and economic reorganization, 52–69
Corruption, campaigns against, and regional rivalry, 214
Costs: and household contracting systems, 36–38, 46, 121; and rural industry, 118; and local leadership, 165; labor, 201–202; and industrial reform, 235, 256
Cotton: and Cultural Revolution, 3; and reform, 10, 83–88; and quota-planning policies, 92–109; and rural income, 114; and industrial reform, 211, 270–278
Crops, commercial, and agricultural reform, 6, 10, 56, 83, 99, 114, 118–120. *See also names of crops*
Cultural Revolution (CR): and reform, 2–10; and centralization, decentralization, 13–14, 25; and Communist Party, 17–18; and education, 19; and incentives, 65, 73–74, 120; and first-category agricultural products, 104; and grain, 119; and household sideline income, 123–124; and industrial reform, 227–228, 253–278
Culture: and reform, 16; and foreign contact, 214

Dalian, and foreign investments, 212
Dazhai system of collectivism, 32–34, 53–61
Decentralization: and agriculture, 10; and industry, 12, 24, 203, 218–221, 223–252, 253, 278; fiscal, 13–14, 25, 223–252, 264, 268; and material allocation, 253–278. *See also* Production responsibility systems
Democratic procedures: and reform, 6; and government, 19; and production responsibility systems, 38–43
Demographics. *See* Population
Deng Xiaoping, and reform, 5–8, 15, 17, 22, 73, 195
Depreciation funds, and industry, 12–13, 226–231, 239, 243–245
Distilleries, liquor, and prices, 199, 249
Distribution: and Dazhai system, 31–34; and agricultural reorganization, 52–53, 61–71, 84; rural income, 125–127; regional, and economy, 198, 210–211
diyi lei wuzi, 85
Doctor, barefoot (non-leading cadre), 27, 163, 171
Dong Qiwu, Vice-Chairman, Chinese People's Political Consultative Conference, and production responsibility systems, 158
Du Runsheng, Director, Rural Policy Research Center, and agricultural reform, 54, 159
Duplication, 14, 265–266, 268–275

Education: and reform, 16, 19–20; and Communist party, 17; and production responsibility systems, 20, 122; and social-change programs, 134, 136, 148, 152; and rural violence, 190
Elderly: and key-point household labor, 46; as welfare provision, 131, 144–145
Electric power: and Cultural Revolution, 3; and industrial reforms, 205; and heavy industry, 208
Employment: off-farm, and Cultural Revolution, 2; and household contracting systems, 37; urban, 125, 268
Energy: and Hua's ten-year plan, 9; and government investment, 201, 220, 247
Environmental protection, and industrial reform, 235
Equality, social, 21–22; and agricultural reform, 160; and industrial reforms, 210
Europe, Eastern, and economic reform, 195. *See also* Hungary
Exchange. *See entries beginning* Market; Trade

Expansion drive. *See* Growth
Export: and reform, 6; and prices, 199-200; and coastal provinces, 212; and localism, 219
Extra-budgetary funds, 26, 245, 278
Extra-plan, outside of plan, 250, 263, 266-268, 277

Fabrics, synthetic and woolen, 249, 271
Family planning. *See* Planning, family
Fans, electric, 249, 272
Farm machinery. *See* Machinery, agricultural
fentian dao hu, 36-38
Fertilizer, chemical: and Cultural Revolution, 3; and agricultural reform, 23, 56; and cotton sales to state, 94; and industrial reform, 260, 270
Feuds, inter-lineage, and local leaders, 16, 175-192
Five Guarantees, and rural families, 133, 151
"Five small industries," and Cultural Revolution, 260, 269
Five Year Plans, First-Fourth: and Cultural Revolution, 4; and industrial reforms, 208, 257-259, 264
Food: and Cultural Revolution, 3; national supplies, and agricultural reform, 24, 57-61; and household contracting systems, 37, 58; and agricultural tax reduction, 117
Foreign investment and trade, 9, 198-205, 210, 217-219. *See also* Export; Import
Four Modernizations: and reform, 5; and Single-Child Family policy, 137-144. *See also* Agriculture; Industry; National Defense; Science-Technology
Fujian province: and production responsibility systems, 40; and local leadership, 162-165; and rural violence, 182-192 and economic reforms, 210-213
Fuzhou, and local investment, 220

Gang of Four: and reform, 6-8, 17-19; and rural violence, 188-189. *See also* Cultural Revolution
gangwei zerenzhi, 169
Gansu province: and production responsibility systems, 40; and local leadership, 167, 170
Glass: and supply shortages, 201; and industrial reform, 271

Goods, consumer: and reform, 8, 14, 25; and state planning procedures, 86; and prices, 199-200, 232; and industry, 205-219, 249-252, 268-275; and rural households, 209
Goods, producer: and reform, 25, 201; and state planning procedures, 86; and economy, 197; and prices, 199-200; and allocation, 257-278
Government: Third National People's Congress, 8, 231; Fourth National People's Congress, 8; Fifth National People's Congress, 8, 207, 214; and reform, 16, 19, 25-27, 161; and Communist Party, 18; and budgetary crisis, 22, 25, 118, 231-232; awards to specialized households, 46; and prices, agricultural, 97-103; State Council Directive No. 51 (birth control), 136; State Statistical Commission, 164; State Planning Commission, 206, 209, 217, 257, 260; State Economic Commission, 206, 213, 217; and population, 139-141; and provinces, 215-216; and government agencies, 216-221; and industrial ministries, 217; and local governments, 197, 218; Sixth National People's Congress, 255; State Capital Construction Commission, 265. *See also* Centralization; Communist Party, Chinese; Governments, local; *entries beginning* Profit; Taxes
Governments, local: and Cultural Revolution, 14; and reform, 14, 24, 204, 218-219; *xiang*, 18; and communes, 18-19, 50-53; finance departments, and local leadership, 24, 228; and agricultural tax reduction, 118; and industrial reform, 14-15, 25-26, 195-252; and materials allocation, 257-278. *See also names of cities, provinces*
Grain, grains: and Cultural Revolution, 3; and reform, 10-11, 57-61, 83-109; and prices, 24; "award," 46; feedgrains, 46, 88, 274; and quota-planning policies, 87-109; rations, and infants, 150; and local leadership, 160; and inland provinces, 212-213
Great Leap Forward, and reform, 5, 25, 73, 259-260
Growth: and modernization, 5-9, 21-23; and agricultural reform, 55; and economy, 197, 200-201
Guangdong province: and commune reorganization, 52-53; and production

responsibility systems, 158-167; and local leadership, 161-173; and rural violence, 180-192; and economic reforms, 210-216, 271
Guizhou province: and production responsiblity systems, 40; and industrial reform, 269

Hainan province, and rural violence, 180, 188
Handicrafts: and agricultural reform, 31; and cotton, 94-96; and rural income, 114; and foreign trade, 205
Hankow (Wuhan), and industrial reforms, 210
Harvests, below-average, and agricultural taxes, 117
Health care: and reform, 16; rural, 26-27; and medical expenses, 27, 122; and social-change programs, 134-156. *See also* Welfare
Hebei province: and education, 20; and production responsibility systems, 40, 158; and household contracting, 47; and local leadership, 171-172; and economic reforms, 210; and material allocation, 266-268
Henan province: and production responsibility systems, 40; and new economic combinations, 49-50; and industrial reform, 274
Hoarding, 265
Households: and production responsibility systems, 35-61, 63-82, 85-86, 120-121, 126; and private exchange, 104; and brigade cadres, 166
Housing: specialized and key-point, 44-47; commodity-grain specialized, 45; and industrial reforms, 202, 248-249, 268
Hu Yaobang, General Secretary, and Party Secretariat, 17
Hua Guofeng, Chairman: and reform, 5-10; and production responsibility systems, 34; and agriculture, 112
Hubei province: and production responsibility systems, 40; and tax-for-profit schemes, 241; and duplicative production, 265-266
Hunan province: and production responsibility systems, 41; and specialized and key-point households, 45; and commune reorganization, 52-53; and rural violence, 181-192
Hundred Flowers movement, and reform, 5

Hungary, and economic reform, 248

Import: food, and Cultural Revolution, 3; equipment, and industrial reforms, 206-207
Incentive schemes: and Cultural Revolution, 3; and socialism, 65; and collectives, 67; and government-industry relations, 14, 25, 195-278; and household contracting, agricultural, 64-65; price, agricultural, 99-101; private trade, agricultural, 105; and Single-Child Family and other social-change legislation, 132-148
Income, family: and Cultural Revolution, 4; and birth control, 20; and production responsibility systems, 122
Income, industrial-commercial, tax, 241-244, 270
Income, peasant, rural, 111-130; and agricultural reform, 2, 10, 11, 22-23, 55-58, 60, 81-84, 138, 178; and contract households, 34-61, 63-69; and state procurement prices, 96-103; and urban income, 111-112; and grain, 114; and local leadership, 157-173; and local violence, 192
Individuals: and production responsibility systems, 63, 120-121; and independent production, 65-68; and private exchange, 104
Industry, 195-278; and reform, 1-16, 22-26; and Hua's ten-year plan, 9; and government, 12; and edible oil crops, 92, and ministries, 216; and decentralization, 219, 261
Industry, heavy: and Cultural Revolution, 3; and reform, 14, 204-210; and "Communist coalition," 16; and economy, 198-199; systems, 207
Industry, rural: and education, 20; and household contracting systems, 49; and economic reorganization, 51-61, 210, 241; and Jiangsu province, 77; and rural income, 114-118, 125; and material allocation, 253-278
Inflation, economic, and industrial reforms, 203, 232
Infrastructure: and Cultural Revolution, 3; and agriculture, 22, 53-61; and collectivization, 72; and coastal provinces, 212; and investment, 220, 268
Inner Mongolia (Nei Monggol) province: and production responsibility systems, 40; and commune reorganiza-

Inner Mongolia *(continued)*
  tion, 52–53; and production brigades and teams, 163
Inputs, agricultural, and production responsibility systems, 121
Inputs, industrial, 201, 253–278
Insurance: labor, 132–133; industrial, 235
Integration, national economic, and interprovincial cooperation, 215, 274
Integration, social-political, and Dazhai system, 32–34
Intellectuals, and Communist Party reform, 18
Interest charges, and enterprise fixed assets, 12, 226, 231, 241–244, 252
International affairs, and reform, 5–6, 211–212
International Rice Research Institute, and rice seed, 3
Inter-regional trade. *See* Trade
Inventories, and material allocation, 266
Investment: and Cultural Revolution, 3; and industrial reform, 2, 11–13, 25, 197–221, 223–233, 245–252; local, 14, 129, 200–201, 219–220, 247–252, 255–278; and agricultural reform, 20–23, 53–61, 122, 128–129, 197; and household contracting system, 37, 47–50; and production responsibility systems, 121–122, 130; fixed capital, and economy, 197, 200; foreign, 198, 204, 219; infrastructure, 212; government, and industry, 200, 209, 219; and materials allocation, 257–278
Iron, iron-and-steel, 206–210, 249, 260–278. *See also* Steel
Irrigation, and production responsibility systems, 122

Jiangsu province: and production responsibility systems, 41, 64, 120; and specialized teams, 72, 77; and economic reform, 210
Jiangxi province: and production responsibility systems, 36–40, 158–159; and private trade, 44
Job security, guaranteed ("iron rice bowl"), and rural reforms, 160, 168

Kin, lineage: and rural violence, 178–192; as labor, and industrial reforms, 202

Labor: division of, and reform, 6, 66; and agriculture, 31, 43–47; and incentives, 67–68; and collectivization, 72, 78; and rural income, 112–113, 120–123; and Single-Child Family policy, 131–156; insurance laws, 132–133; rural, and reforms, 202; and industrial reform, 201–202, 208, 226, 231
Land: and reform, 23–24, 58; contract to households, 31–62; and surplus labor, 43–47; fodder, to specialized households, 46; and collectivization, 72–82; and production responsibility systems, 121; and local leadership, 169–170; reform, and rural violence, 177–192
Leadership, local, agricultural, 2, 16, 60–61; and family planning, 150–156; leading cadres, and rural reforms, 157–173; non-production, 167–173; and rural violence, 178–179, 185–192; duties of, 186; and industrial reforms, 208
Leadership, local, industrial, 23; and agricultural local leadership, 169–170; and industrial reforms, 208, 218–221, 275
Leadership, state: and Communist Party reform, 19; and rural violence, 175–192
Legal system: and reform, 6, 19; contract law, 19; criminal code, 19, 182; and labor insurance, 132–133; and marriage, 132–133, 136, 143–146
lianchan jichou, 34–36
Liaoning province: and grain quotas, 88; and industrial reform, 236, 256, 266–268
li gai shui, 217, 240
Li Kaixin, Director, State Materials Bureau, 266–268
Liquor. *See* Distilleries, liquor
Livestock-raising, 43; and production responsibility systems, 121
Living standards: and Cultural Revolution, 4; and reform, 6–8; rural, 11, 23, 60; and specialized housing, 46
Li Xiannian, 8
Loans. *See* Bank loans, repayable
Localism, and industrial reform, 219–221, 238, 274
Location, geographic, and specialization, 196
Looting, and rural violence, 187

Lumber, and industrial reform, 250, 255, 262-268

Machinery, agricultural: and Cultural Revolution, 3; and agriculture, 23; and household contracting, 47-49, 209; and industrial reform, 260, 270-278
Machinery, industrial: and industrial growth, 201; and foreign trade, 205-206; and industrial reform, 207-210, 219, 248-252
Managerial systems, rural, and reform, 43-61, 63
Manufacturing. See Industry
Mao Zedong, Chairman, and reform, 5-6. See also Cultural Revolution
Market, marketing: and agriculture, 2, 10-11, 31, 44-61, 84-109, 118, 212-213; independent, and industry, 12, 25, 225-252, 266-278; and local governments, 14; state control over commerce, 57-61; area, and commune, 178; international, and light industry, 205; and localism, 219
Markets, farm commodity: and agricultural reform, 31; urban, 45; and household sidelines, 45, 123-125; and agricultural products, 104-109; and grains, 104
Material allocations, and industry, 2, 12-14, 25, 229, 250, 253-278
Materials, building, construction, 201, 203, 219-220, 249, 269-278
Materials Bureau, State, 254-255, 266-268
Materials, raw: and prices, 199; and industrial reform, 248, 252
Materials exchange fairs, 266
Metals, nonferrous: and foreign trade, 206; and industrial reform, 208-210, 262-268
Middlemen, and agricultural private exchange, 104
Migration, restrictions against, and agriculture, 177-178
Military objectives. See National Defense
Ministries, bureaux, industrial: and "Communist coalition," 16; and government reform, 19, 217-218; and economy, 197-200, 216, 239, 245, 257, 260-261
Ministry of Agriculture, 65-82
Ministry of Building Materials, 276

Ministry of Finance, 21, 217, 229-230
Ministry of Foreign Economic Relations and Trade (MOFERT), 209, 217-218
Ministry of Geology, 217
Ministry of Light Industries, 209
Ministry of Machine Building (MMB), 206, 209, 234
Ministry of Metallurgy, 206
Ministry of Petroleum, 206, 217
Ministry of Textiles, 209
Minorities: and family planning, 147; and religion, 183-184
Morale, worker, 4, 234
Motivation, and economic organization, 66-68

National Defense: and Single-Child Family policy, 132; and industrial reforms, 210, 262

Oil exploration, offshore, 217
Oils, edible: and Cultural Revolution, 3; and reform, 10; and quota-planning policies, 83-109; and rural income, 114
Open-door policy. See Trade, foreign
Opium, outlawing of, and social change, 133
Output, agricultural: and production responsibility systems, 10, 36-38, 63-64; and technical responsibility systems, 48; and reorganization, 55, 68-69
Output, industrial: and Cultural Revolution, 3; and industrial reform, 2, 12, 199-200, 233; and material allocation, 261
Overhead, and productivity, agricultural, 168-171

Peasants, and Communist Party reform, 18. See also Agriculture; Leadership, local, agricultural
Penalties: and household contracting systems, 36-38; and Single-Child Family campaign, 132, 141, 149
People's Communes, and reform, 5, 18
People's Daily: report on Cultural Revolution and economics, 4; and Communist Party reform, 18; and household contracting systems, 39; and Single-Child Family policy, 131-

People's Daily (continued)
132; and rural production responsibility systems, 158; and local leadership, 172; and rural violence, 181–192; and industrial reform, 230, 235
Petroleum industry: and foreign trade, 206; and industrial reform, 207–210
Pharmaceuticals, 270
Phosphorus ores, 269
Planning, agricultural: and collectives, 24, 76–77; and reform, 54–61, 84–86; production and quota, 99–109
Planning, economic: and "Communist coalition," 16; and trade, 14; and reform, 197–221; and material allocation, 262–263
Planning, family: and agriculture, 2; and reform, 16, 20–21, 26–27, 58; and production responsibility system, 20; and Single-Child Family policy, 131–156
Politics: and industry, 2, 207–210; domestic, and reform, 5–10, 14–15; and economics in reform, 2, 15–21, 31, 51–61; and educational reform, 20; and demand for commodities, 127; and behavior, 161; and economic reform, 195–197; and inflation, 203
Poor or mountainous areas: and production responsibility teams, 63–64, 77; and grain, 87; and Single-Child Family policy, 146
Population: growth, during Cultural Revolution, 2; and agriculture, 10, 24, 60; and Single-Child Family policy, 138–139
Ports, and industrial reform, 211
Prices: and agriculture, 10, 22, 56, 85, 96–103, 114–117, 121, 123–130; and industry, 12, 14, 24–25, 299–200, 232, 234–235, 241, 249–252, 268–278; food, 22; and indexes, procurement and sales, 98–99; and planning, 99–101, 197; and inflation, 203. *See also* Prices, above-quota; Prices, negotiated; Prices, quota
Prices, above-quota: and agriculture, 10–11, 23, 99–109; and grain, 23; and rural income, 114–116
Prices, free-market. *See* Market, independent, and industry; Markets, farm commodity; Private sales, outside-of-plan
Prices, negotiated: and rural income, 114–117; and industrial reform, 268–278

Prices, quota: and agriculture, 10–11, 22–23, 99–109; and cotton, 97–103; and grain, 24, 97–103; and oil, edible, 97–103; peanut, 99; rapeseed, 99; and rural income, 114
Private exchange. *See* Market, independent, and industry; Markets, farm commodity; Private ownership, plots; Private sales, outside-of-plan; Trade, private
Private ownership, plots: and agriculture, 11–12, 23–24, 31, 37, 43–47, 56, 114, 123–125; and collective farming, 76–82, 121, 138; and production responsibility systems, 123–124; and rural income, 125
Private sales, outside-of-plan, and industrial reform, 225–252; 266–278
Processing industries: and local investment, 14; and prices, 199; and heavy industry, 205, 219
Producer goods. *See* Goods, producer
Production, productivity: and agricultural reform, 2, 6, 44, 50–61, 64–82, 83–84, 83–109; and household contracting, 46; and rural income, 114–116; and economy, 197; and cost of labor, 202; and profit retention, 226–227, 231; and materials allocation, 260–278
Production responsibility systems (PRS): and agriculture, 10, 14–24, 34–62; and Single-Child Family policy, 137–156; and local leaders, 163–173, 179; and heavy industry, 209; and coastal and inland provinces, 212–213
Profit contract systems, and industry, 11–12, 24, 233–240
Profit retention, enterprise, and industry, 11–14, 203–210, 226–233, 239, 244–245, 268
Profits: and taxes, 11–13, 26, 240–244; above-quota, 12; and economy, 197–200; and bonuses, 202; and industrial reforms, 203–204, 205, 208, 217, 219; and localism, 220; and profit retention, 227, 249–252; and material allocations, 269–278. *See also* Profit retention, enterprise
Profit-sharing. *See* Profit contract systems; Profit retention, enterprise
Propaganda: and social-change policies, 132, 134; and rural violence, 190
Property, and reform, 178
Prostitution, outlawing of, and social change, 133–134

Protectionism, 273-275
Provinces, coastal, and industrial reforms, 204, 210-214
Provinces, inland: and "Communist coalition," 16, 204; and industrial reforms, 214-216

Qinghai province, and religion, 183
Quotas, above-quota: and grain, 87-91; and agricultural products, 99; and rural income, 116, 130
Quotas, labor, and field work, 120
Quotas, procurement: and agriculture, 10, 85-109; and household contracting systems, 37, 85; and rural income, 115; and economy, 197. *See also baochan daohu*
Quotas, profit-remittance, and industry, 12
Quotas, work, and industrial reform, 239-240

Railroad, and local investment, 220
Rectification, enterprise, and industrial reform, 239-240
Regions: and industry, 6, 14, 203-204; and agriculture, 10, 23, 54-61; and quotas, 85-86; and trade, 10, 266, 274; and economy, 198; and economic conflicts, 210-216; and material allocation, 257-278
Religion, superstition: and local leaders, 16; and rural violence, 178-192
Remuneration: worker, and industry, 11; Party cadre, and reform, 16; and production responsibility teams, 34-61; and technical responsibility systems, 48; and commune reorganization, 52; and labor, 66; and local leadership, 168-171. *See also* Wages
Renting, tenancy: of land, 23, 76-82; and production responsibility systems, 121, 130
Repair work, and agriculture, 44
Research institutes, and educational reform, 20
Resistance to reform: and agriculture, 15, 28-29, 46-47, 57-58, 175-192; and industry, 16; and local leaders, 16; and Single-Child Family policy, 131-133, 153
Resource, resources: allocation, and industrial reform, 2, 227-228, 239, 244-252, 253-278; and Cultural Revolution, 4; and agricultural reform, 61, 160; and production responsibility systems, 122; and rural violence, 177, 181-192
Responsibility systems, economic. *See* Profit contract systems
Revenue-sharing, 25, 268
Rice: seed, and Cultural Revolution, 3; and private exchange, 105; and agricultural tax reduction, 117-118
Roads, and local investment, 220
Rural sector. *See* Agriculture: *specific categories*

Sales, selling: and agricultural reform, 31, 104; and rural income, 114, 126; private, 225-252, 266-278. *See also* Income; Prices
Scale, economies of, and collectives, 71-75
Schools, key-point, and educational reform, 20
Seed technology, and Cultural Revolution, 3
Self-sufficiency: and food production, 3; and reform, 6, 9, 10; and Dazhai system, 33-34; and agricultural reform, 55, 84; and grain production, 85-91; and edible oils, 92; and regions, 262-265
Shandong province: and household contracting systems, 39-43; and production brigades, teams, 163-165; and rural violence, 187; and industrial reform, 234-236
Shanghai: and pork rationing, 3; and household contracting, 64; and specialized teams, 72; and industry, 199, 210-216, 219, 230, 242, 272-278
Shanghai Watch Factory, and industrial reform, 269-272
Shaanxi province, and private trade, 44
Shanxi province: and production responsibility systems, 41, 158; and water conservancy management, 48; and local leadership, 166; and Chinese culture, 214; and barter trade, 263-264, 274; and coal production, 269, 274
Shortages: and Cultural Revolution, 3-4; and material allocation, 266
Sichuan province: and agriculture, 11-12, 213; and industry, 11-12, 14, 226, 230, 242; and commune reorganization, 50-53, 154; and collectivization,

Sichuan province *(continued)* 80; and household sideline income, 123-124

Sideline production, agricultural, 11-12, 23-24, 31, 37, 43-47, 56, 114, 123-125

Single-Child Family policy, 20-21, 131-156

Social services. *See* Welfare

Soviet Union: and internal reform, 6; and collectives, 67; and economic reform, 195-197; and materials allocation, 257

Special Economic Zones (SEZs): and foreign investors, 212; and Special Economic Zone Office, 213-215

Specialization: production and marketing, agricultural, 35-36, 46, 50-61, 103-109, 128; bureaucratic, and economic reform, 196-197; and regions, 210-216, 253, 278

State-owned enterprises: and Cultural Revolution, 4; and economic reorganization, 52-53; and agriculture, 85-109; and rural violence, 187. *See also* Profit contracting; Profit retention

Steel: and reform, 201, 250, 255, 260-268; and foreign trade, 205-206

Subsidy, government: and food, 22, 57-61; and local leadership, 167-171

Sugar industry: and Cultural Revolution, 3; and prices, 199; and industrial reform, 272-273

Superstition. *See* Religion, superstition

Supply and demand: and Cultural Revolution, 3; and agriculture, 23; and material allocation, 269, 275

Supply and Marketing Cooperative, state-run, and commune reorganization, 52

Supplies, and economy, 197, 219

Taxes: and agriculture, 11, 79-82, 99; and industry, 12-13, 24-25, 197-198, 203-204, 208, 217, 249; and local investment, 14, 200-201, 220; and education, 20; and household contracting systems, 37; and grain, 87-91, 115, 117-120; and communes and brigades, 116-117, 126; multichild, 149; for communal social services, 151; likin, 211-212; income, 241-244, 270-278. *See also* Tax-for-profit schemes

Tax-for-profit schemes, and industry, 11-13, 26, 240-244

Team: and collective ownership, 33; mechanized, and technical responsibility systems, 48; and new economic combinations, agricultural, 50

Teams, production: and production responsibility systems, 34-61, 65-69, 120-123, 126; and commune reorganization, 52-53; and incentives, 67; specialized, 72; and rural income, 112-122; leaders, and reforms, 160-161, 157-173; and kinship, 178; and rural violence, 178-192

Technology, and economic organization, 65-68

Technology, agricultural: and household contracts, 47-50; and agricultural reorganization, 52-53, 61

Technical responsibility systems: and household contracting, 48-49

Ten-year plan, Hua Guofeng's, 8-9

Textile industry: and cotton, 94-96; and rural income, 116; and foreign trade, 205; and industrial profits, 211; and synthetic and woolen fabrics, 249; and prices, 272

Theft, of state property, and rural violence, 187

Tianjin, and industry, 199, 210-216, 219, 230, 236, 242, 270-271

Tie Ying, governor of Zhejiang, and production responsibility system, 158

tonggou tongxiao, 85

tongguan, 257

tongyi jingying, lianchan jichou, 35-36

Tools: and agriculture, 23; and production responsibility systems, 121; and industrial reform, 249

Tractor drivers: and rural income, 81; and rural violence, 187

Trade: and regions, 10, 263, 266, 274; and specialized and key-point households, 44-47; foreign, 198-200, 204-205, 210, 217-218; barter, 215, 263, 274. *See also* Markets, and agriculture; Markets, farm commodity

Transport, transportation: and Cultural Revolution, 3; and Hua's ten-year plan, 9; and government investment, 201, 220, 247; and local investment, 220

Trucks and tractors: and private ownership, 45; and industrial reform, 262-264

United States: normalizing relationship with, 6; and internal reform, 7; and economic reform, 196; and industrial reform, 210
Urban areas: and household contracting, 64; and grain, 87; and edible oils, 92; and income and consumption, 111-112, 126-127; and private marketing, 124; and employment, 124, 268
Urban sector. *See* Industry

Vegetable oils. *See* Oils, edible
Village, administrative: and commune reorganization, 51; and social organization, 177; and brigade, 178
Violence, rural: and agricultural reform, 2, 26, 175-192; and local leaders, 16

Wages: for state workers, and Cultural Revolution, 4; and economy, 197; and industrial reforms, 201-202, 208, 235, 245
waixiao, 274
Wan Li, Governor of Anhui Province, and production responsibility systems, 120
Wasteland, and agriculture, 44
Watchmaking, and industrial reform, 270-278
Water, and rural violence, 179
Water-conservancy management responsibility systems, and contract households, 48-49
Water conservancy projects, and Cultural Revolution, 3
Welfare provisions: and reform, 7-8, 11; and government investment, 20; and family planning, 20-21, 132-156; and agricultural reform, 24; and Dazhai system, 33-34; and household contracting systems, 37; and incentives, 66; and production responsibility systems, 122; and social change, 132-135; and social assistance programs, 133, 142, 144; grain payments for, 170; and profit retention, 227
Wheat, and private exchange, 105
Wholesale, resale, and agricultural private exchange, 104
Wine-making, 272-273
Women: and key-point household labor, 46; and collectivization, 80; and agricultural labor, 144
Woodlands, and rural violence, 179-192
Work points, payment in: and technical responsibility systems, 48; and collectives, 67, 74, 138; and production teams, 68-69; and local leadership, 169
Wuhan Iron and Steel Company, and local leadership, 208
Wuhan, and industrial reforms, 210, 213

Xue Muqiao, economist: and regional distribution, 214; and taxes, 272

Yao Yilin, Vice-Premier, and economic reform, 255
Ye Jianying, and Deng Xiaoping, 8
Young workers, and Cultural Revolution, 4, 268
Yuan Baohua, State Economic Commission, and responsibility systems, 239
Yunnan province, and industrial reform, 269

Zhang Jingfu, Finance Minister, and planning group, 228-230
Zhao Ziyang, Premier, and agriculture, 108, 116, 138
Zhejiang province: and production responsibility systems, 37, 41, 64, 158; and rural violence, 181-182; and economic reforms, 210
Zhou Enlai, Premier: and reform, 6, 8; and birth control, 136
zhuanye chengbao, lianchan jichou, 35-36
zhuanye hu he zhongdian hu, 44-47